Studies in Computational Intelligence

428

Editor-in-Chief

Prof. Janusz Kacprzyk
Systems Research Institute
Polish Academy of Sciences
ul. Newelska 6
01-447 Warsaw
Poland
E-mail: kacprzyk@ibspan.waw.pl

For further volumes:
http://www.springer.com/series/7092

Johannes F. Knabe

Computational Genetic Regulatory Networks: Evolvable, Self-organizing Systems

Springer

Author
Johannes F. Knabe
Science and Technology Research Institute
University of Hertfordshire
Hatfield
United Kingdom

ISSN 1860-949X   e-ISSN 1860-9503
ISBN 978-3-642-30295-4   e-ISBN 978-3-642-30296-1
DOI 10.1007/978-3-642-30296-1
Springer Heidelberg New York Dordrecht London

Library of Congress Control Number: 2012938579

Foreword

Genetic Regulatory Networks (GRNs) in biological organisms are primary engines for cells to enact their engagements with environments, via incessant, continually active coupling. In differentiated multicellular organisms, tremendous complexity has arisen in the course of evolution of life on earth. In some organisms, such as humans, this occurs with on the order of 10^{13} to 10^{14} cells of hundreds of different types largely cooperating in complex structures to give rise to reliably inheritable organization and form, growing from a single cell and remaining viable at each stage. Engineering and science have so far achieved no working system that can compare with this complexity, depth and scope of organization. Levels of complexity are routinely handled by biological systems, which still seem an insurmountable challenge to current practices in system engineering and computation at present. We currently do not yet know how to 'grow' such complex engineering and computational systems. Thus it makes good sense from a biological and also from computational and engineering viewpoints to examine how this is possible in nature with its wonderfully sophisticated and successful living differentiated multicellular systems.

Abstracting the dynamics of genetic regulatory control to a computational framework in which artificial GRNs in artificial simulated cells differentiate while connected in a changing topology, it is possible to apply Darwinian evolution in silico to study the capacity of such developmental/differentiated genetic regulatory networks (DGRNs) to evolve. Pioneers such as Dr. Johannes F. Knabe have been leading the way in doing just that and more. This serves not only as an abstract model for evolutionary developmental biology, but as a novel paradigm for computation, and the evolutionary organization of complex environmental embedded processes.

In this volume, J. F. Knabe introduces a GRN model in which cells' internal environment comprises concentration levels of various proteins, including those that can be produced by the GRN, such as transcription factors (TFs) which regulate the GRNs activity. Cell state is also influenced by external signals (such as morphogens that diffuse from nearby cells). The cell's activity orchestrated via its individual GRN controls its dynamical behaviour, with different cell types corresponding to different attractors of the genetic regulatory network considered as a dynamical system. Variants of this evolutionary GRN paradigm are here investigated for their evolvability and robustness in models of biological clocks, of simple differentiated multicellularity, and of evolving artificial developing 'organisms' which grow and express an ontogeny starting from a single cell interacting with its environment, eventually including a changing local neighbourhood of other cells. Examples of the latter hinting at the power of this approach use (Glazier-Graner differential adhesion) cell-sorting dynamics as part of the dynamic environment to evolve multicellular DGRNs that autonomously set up morphogen gradients and solve Lewis Wolpert's canonical problem of evolving a 'French flag', i.e. an organism that grows from a single cell into an differentiated multicellular form expressing 'blue', 'white' and 'red' cells in vertical stripes.

Such evolved ontogenically responsive and robust patterning can of course serve as the substrate for further growth and differentiation as in real living things, but potentially also for computational tasks in which resources can be specialized, adapted, and deployed – like limb buds growing into elaborated new structures away from the present scope of the body – in a massively parallel, self-structuring manner to solve new, unseen instances of a problem, or new types of problems. These methods may help us understand the genesis, organization, adaptive plasticity, and evolvability of differentiated biological systems, and may also provide a paradigm for transferring these principles of biology's success to computational and engineering challenges at a scale not previously conceivable.

Hatfield, England, March 2012   Chrystopher L. Nehaniv

Acknowledgements

Most of the research presented here was conducted while the author was at the Adaptive Systems Research Group, School of Computer Science, University of Hertfordshire. So first of all I would like to thank my PhD supervisors, Chrystopher Nehaniv and Maria Schilstra. You would not read this book without their invaluable support, infectious enthusiasm and willingness to share their vast experience.

I would also like to thank my colleagues in the Adaptive Systems and Neural Networks & BioComputation research groups. They were always available for fruitful discussions and the occasional conference trip. The CS technicians were most cooperative in setting up a Condor computing cluster on the student lab PCs, running jobs from our group (including most experiments reported here) when the machines were idle.

In addition to the above, I want to thank all the passionate and intelligent individuals, teachers and peers, who shaped my mind, intellect and personality, in school, in the Cognitive Science degree at the University of Osnabrück, Computer Science at the Norwegian University of Science & Technology and the InDy group of the Fraunhofer Institute, and, last but not least, outside academic context.

Very special thanks to my family and friends for being patient and the best backing I could ask for.

Contents

Chapter 1
Introduction

Everywhere on earth life abounds. Species have been able to come up with solutions to "engineering problems" in innovative ways over and over again to advance into and survive in all kinds of environments, sometimes more constructing than discovering their very special ecological niches where one would not expect any. Species are made up of myriad individuals which in turn consist of one or many cells. Cells are responsive to external signals, allowing them to cope with changing conditions and perturbations, while in multicellular organisms cells autonomously "negotiate" division of labour among them. Naturally this adaptability and diversity has fascinated and continues to fascinate humans.

A better understanding of how organisms function and how they change over generations promises not only insights into human biology and evolution but also advances for our technology: Nature has long been a source of inspiration for engineers – think of the hook-and-loop fastener and burdocks, to name a prominent example of bio-mimicry or, closer to this research, take artificial neural networks or [Beer(1994)] Viable System Model for organisational structures, inspired by the adaptive self-organisation of the brain. At a time where industrial processes are getting more complex and variable, ideas from biology can improve the automated design of complex systems and lead to novel computation paradigms.

Investigating the basic principles responsible for organismic control, organisation and evolution is the topic of this research.

Evolution (adaptation, heredity, and variation) certainly is a powerful principle behind much biological adaptability. However, evolution-inspired algorithms have shown that evolution is only as good as what it can work with: a lack in evolvability, "the ability to reach 'good' solutions via evolution" [Nehaniv(2003)] (definition modified from [Altenberg(1994)]), has restricted their success. Evolvability crucially depends on the nature and variational properties of the genetic material and how this translates to selectable, hereditary traits[1]. Nature's solutions, gene networks and multicellular development, apparently do provide enough evolvability, but why do they work so well?

---

[1] See section 2.4 for a more formal discussion.

J.F. Knabe: Computational GRNs: Evolvable, Self-organizing Systems, SCI 428, pp. 1–5.
springerlink.com   © Springer-Verlag Berlin Heidelberg 2013

A first hint is given by the fact that there are not enough genes for a construction program detailing every single cell's exact position and function for a whole human. Our organism consists of ten trillion cells (of which about a hundred billion are neurons in our brain). They all harbour the same genome[2], so cellular communication and self-organisation have to play a role. Furthermore, gene number is not an indicator of (perceived) complexity of organisms. The human genome of three billion letters or base pairs codes for roughly 25,000 genes (more than 90% of genetic material is not directly encoding protein) – not much more than in other organisms, e.g. mice also have about 25,000 genes. There is a high degree of similarity between different species, for example roughly 99% between chimpanzees and humans.[3] Most differences between related species are in the timing, strength or spatial location of gene expression, rather than in the function of the proteins that the genes encode [Carroll(2008), Carroll et al(2001)Carroll, Grenier, and Weatherbee]. Even small changes in gene expression patterns can have a big impact: The proteome[4] is, unlike the relatively static genome, highly dynamic. Not only are proteins continuously synthesised and broken down. Their distribution also changes all the time due to regulatory processes: The product of one gene can regulate the expression of others or even itself. Direct or indirect self regulation (feedback loops) may result in very complex dynamics. Structurally however these interactions can conveniently be depicted as so-called genetic regulatory networks (GRNs) – with the genes as nodes and connections where regulation takes place between genes. Often regulation is additionally influenced by external stimuli. Such stimuli can be signals from other cells or environmental, like the presence or absence of a nutrient. The former is likely in colonial or multi-cellular, the latter in single-celled organisms. Gene regulation allows for the same gene to be expressed differently, in a context sensitive way – important for gene re-use and responsiveness to the environment. On a larger scale, gene regulation allows for cellular differentiation, the basis of division of labour between the cells of an individual (there are more than 200 cell types in a human).

So there is a generative, non-linear mapping from hereditary material to organism, with interaction at all levels of organisation – involving the interplay of physical environment (embodiment), intercellular communication and intracellular gene regulation dynamics [Maynard Smith(1998)].

---

[2] The genome contains of bases and holds the molecular information for building and maintaining an organism. These 'instructions' are continuously read. Special areas on the genome, the genes, encode proteins – the cell's building blocks, workhorses, and messengers.

[3] Note that all the numbers mentioned in this paragraph are estimates; values reported in the literature vary sometimes by one order of magnitude. Good starting points are, for example, [Stein(2004)] and [Margulis and Sagan(1997)].

[4] The proteome is the entirety of proteins expressed in a particular cell at a particular time. For humans the overall set is roughly 100,000 proteins, much higher than the number of coding genes due to alternative splicing and other modifications – see section 3.1.

This has supported the evolution of modular development and repetition. Bilateral symmetry is only one example, others include vertebrae or segments in insects like centipedes. Such re-use opens the door to repetition with modification, as for example legs are not the worst basis for antennae (see later chapters and especially fig. 2.2 for details). In nature, "solutions" are also generalised to different scales or proportions. An example for this scalability is that mouse and humans have a similar number of genes but the number of cells they are made of differs by orders of magnitude.

Evolutionary algorithms also use encodings that define potential solutions to a problem. Contrary to the above, they have typically used direct mappings from hereditary material to solutions[5]. For example one "gene" could encode the numeric value of one parameter of a function to optimise. While this is computationally efficient and serves the purpose of optimising simple functions we cannot expect complex, surprising solutions to arise. The problem has been realised by researchers and efforts have been made to improve evolutionary algorithms. Many attempts of including a more dynamic, generative mapping however are still very restricted, using a number of complex building blocks with little interaction between them (or environmental input) instead of simple, uniform ones. In addition, such construction codes are generally disposed of after the mapping, while biological GRNs play an important role also during an organism's life.

As a result, traditional evolutionary algorithms do not achieve very high evolvability, scalability and adaptability.

1.1 Research Questions and Aims

Given the problem of limited adaptability and evolvability exhibited by conventional evolutionary algorithms, and that the solution is using non-linear encodings akin to those successful in nature, this work is motivated by the following hypothesis:
Appropriate computational models of genetic regulatory networks can exhibit a high degree of evolvability.

Central research questions are defined, on the grounds that answers to them will provide evidence towards this hypothesis.

1. Do computational models of GRNs have characteristics such as adaptability and robustness in common with their biological counterparts?
2. Is it possible to evolve artificial GRNs with crucial functionality, such as ability to oscillate and differentiate, from random initial configurations and what influences their evolvability?
3. Does local network topology reflect evolved function such as switching, i.e. are certain connectivity patterns associated with dynamics?
4. How unique are the networks that realise particular functionalities and how robust are they?

---

[5] This fact might actually be not too surprising as evolutionary biology also largely ignored (or was not aware of) the importance of genetic interaction and development.

5. Can models of dividing cells, each daughter with the same GRN controller and restricted information only, be evolved to self-organise into differentiated patterns?
6. What is the influence of embodiment and thus physical constraints?

1.2 Methodology

As mentioned, computational modelling and simulations are used to investigate the questions listed above. The main motivation for this research is that harnessing the evolvability of natural GRNs could lead to huge improvements in automatic engineering. However, it is also hoped that results can feed back into biology in the following sense: This work does not aim to achieve a one-to-one correspondence with real life organisms but create systems which model the core natural processes to gain a better understanding of them. The target is not only to reproduce global macro level phenomena from nature but to do so using the key lower level structural processes from biology. Such a "bottom up" constructive approach, situated in the realm of "Artificial Life", can capture qualitative behaviours of overarching interest and help ask new questions.

In order to establish this work's hypothesis, it suffices to demonstrate the existence of one or more computational GRN models that exhibit high evolvability. Identifying "appropriate" ingredients for such a model must comprise part of this investigation, as must the employment of measures of evolvability. In fact, several such computational models and measures of evolutionary capacity are discussed.

The GRN models developed in this work (section 3.3) build on and extend [Quick et al(2003)Quick, Nehaniv, Dautenhahn, and Roberts] *BioSys* model, as it has been shown a powerful basis for embodied active control systems (see section 3.2).

This research also employs (and extends) [Alon(2006)] topological motif analysis (chapter 5) and adapts the Cellular Potts Model [Glazier and Graner(1993)] for a simulation of physics (section 6.3.1), due to its good balance between computational tractability / abstraction and realism with respect to the laws of physical space.

1.3 Contributions

The main contributions to knowledge presented in this work are:

1. The introduction of a genetic regulatory network model that uses principles overlooked or considered irrelevant before (chapter 3).
2. A demonstration that the model shows properties of their natural counterparts, like adaptation, robust control and differentiation (chapter 4).
3. The identification of mechanisms that systematically influence the model's evolvability (mostly chapter 4).
4. Providing evidence that topological sub-graph pattern analysis is a poor basis for drawing conclusions about the dynamics of the system (chapter 5).

5. The development of a physical multicellular simulation with a good balance of computational tractability and abstraction (chapter 6).
6. A demonstration that with such a model it is possible to evolve a computational analogue of biological morphogenesis: embodied, self-organised pattern generation (chapter 6).

1.4 Overview of Chapters

This book is structured as follows:

In chapter 2 an introduction to evolution is given, first reviewing the natural process, including concepts from developmental biology. Then algorithmic abstractions of evolution are discussed. As an understanding of evolvability is important for this work there will be special focus on what it is, how it is realised in nature and why there often is a lack of it in models.

Genes, as well as the regulatory actions between them control cell behaviour during development but also the lifetime of most cells. They are a major subject here, so their functioning is introduced in depth in chapter 3. Based on this, models of GRNs are reviewed. After the overview, particulars of the basic GRN model, xBioSys, and evolutionary setup are described.

Chapter 4 describes two important cell behaviours, namely "timekeeping" and differentiation. Why these behaviours were chosen as representative benchmarks is explained and initial experiments carried out. Afterwards results and implications from the experiments are presented.

In chapter 5 network analysis methods are investigated, specifically "motif" analysis. This method seemed a promising candidate for breaking down GRN complexity and detect small patterns with particular functionalities. However, things turn out to be more complicated. Based on these results arguments suggesting that some results previously produced with the method were based on flawed assumptions are presented.

Morphogenesis, the process that causes an organism to develop its particular shape, is the focus of chapter 6. After a short review of the physical dimension of the biological process, a spatial physical model (the so-called Cellular Potts Model) that allows arbitrary cell shapes while still being computationally tractable is presented. In experiments with the model, self-organised differentiation (as opposed to externally induced differentiation) for spatial patterning and scalability is found.

Finally a review and discussion of what was learned from this research, and possible directions for future work, conclude the book.

Chapter 2
Evolution

Evolution created the huge diversity of species we see on earth nowadays, each adapted to their habitats. Of course their evolutionary history also shaped organisms, prompting [Dobzhansky(1973)] to utter his famous: "Nothing in biology makes sense except in the light of evolution". A very general definition of evolution in nature is "change in the gene pool over time". (Where the 'gene' is a hereditary unit that can be passed on unaltered for many generations.[1] The gene pool is the set of all genes in a set of several individuals or population.)

An individual, a single member of a population, has a genotype and a phenotype. The genotype is the collectivity of all genes (in the sense above), while the phenotype is the expression of that genotype in an environment. When resources are limited, phenotypes that are better adapted to the environment have a tendency to produce more offspring that survives to reproductive age – they are said to have a higher fitness. Mutation and recombination cause changes to the genome in a random manner, possibly affecting the carrier's phenotype. Such phenotypic variation, possibly affecting fitness, is the raw material for selection. As the genotypic variations arc heritable, the proportion of carriers of the advantageous genes in the population increases over time.

The population is evolving over generations - iterations of evaluation, selection and reproduction with variation. This can be depicted as evolutionary cycle (see fig. 2.1): Parents beget offspring, which receive a recombination of parental genes, possibly changed. The young replace the old in the population, while selection favours fit individuals to be parents. This basic cycle is the principal idea underlying evolutionary computation, described in section 2.3. A central topic of this work, evolvability, "the capacity of a population to generate adaptive heritable genotypic and phenotypic variation" [Nehaniv(2005)], is reviewed in the last section of this chapter. The following section however addresses particulars of natural evolution (mostly concepts related to this research, as a full account is beyond the scope of this work), some of which are often neglected in computational abstractions.

---

[1] Note that this use of 'gene' as informational unit differs from the bio-physical meaning of the word.

J.F. Knabe: Computational GRNs: Evolvable, Self-organizing Systems, SCI 428, pp. 7–18.
springerlink.com

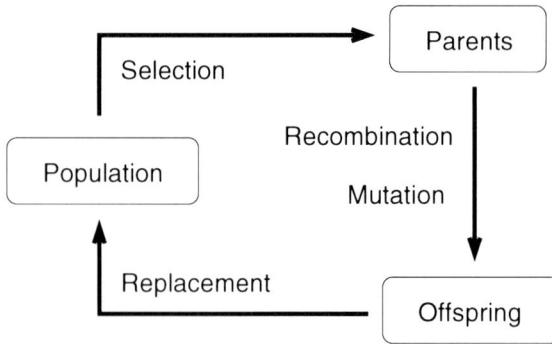

Fig. 2.1 Schema of the basic evolutionary cycle

2.1 Biological Evolution

The theory of natural evolution was first presented in detail by [Darwin(1859)], but it took more than 50 years to incorporate [Mendel(1866)] rules of molecular inheritance, population genetics and other insights to create the so-called modern evolutionary synthesis [Huxley(1942)]. While details are still being added and the relative influences of mechanisms are debated, the modern synthesis ("Neo-Darwinism") is widely accepted by scientists.

Mendel studied the inheritance of discrete traits, such as the violet or white colour of pea blossoms, which he found to be dominant or recessive – as opposed to "blending inheritance" where an offspring was thought to be the average of its parents. This, together with the insight that often several loci control non-discrete traits[2], and [Morgan et al(1915)Morgan, Sturtevant, Muller, and Bridges] chromosome theory of inheritance, classical genetics was established. Another major insight was provided by population genetics: the importance of stochastic effects. Due to sampling errors certain variants of traits may become more common in a population, even if there is no (immediate) selective advantage. This genetic drift was first described by Wright in the 1920s[3]. Later [Kimura(1983)] formulated neutral evolution as a special case of genetic drift, noting that many mutations are neutral or near neutral to the phenotype. While some genome regions are highly conserved over evolutionary time as any change to them would be fatal, selectively neutral parts of the genome change at a constant background mutation rate, providing a molecular clock that is often used to estimate the time which has passed since a split into two species occurred.

While individual stochastic effects are usually small, accumulated events can lead to the irreversible loss of a variant. Especially in small or isolated populations the sampling error gets bigger. For example, genetic drift along with small

---

[2] Relatively few traits are influenced by only a single gene, for example Huntington's disease in humans.

[3] After him it is also referred to as "Sewall Wright effect".

population size is suspected to play an important role in speciation: In spatial isolation (e.g. on an island) parts of a population might accumulate stochastic effects up to a point where the progeny can no longer interbreed with the main population – the split into two species.

The fact that populations are distributed in space, and mating (as well as other interaction, e.g. competition) is mostly localised also plays an important role in maintaining diversity.

Evolution has gone through major transitions, where previously unknown levels of complexity and organisation were achieved. One such changeover in evolution was the Eukaryote [Maynard Smith and Szathmáry(1998)]. Unlike Prokaryotes (Eubacteria and Archaea), Eukaryotes (animals, plants, Fungi, and Protists) have complex internal organisation with membrane-enclosed sub-structures. The largest is the nucleus, which holds (most) genetic information. Another transition in evolution was the advent of multicellularity[4] (for a discussion of its origin and the advantages of reproduction via single egg cells followed by morphogenesis see [Wolpert and Szathmáry(2002)]). Due to development and mechanisms building on it, variation at the protein level can map in very non-linear ways to phenotypic variation. In some cases potential variability is reduced and canalised into a fairly invariant phenotype, due to developmental buffering or "canalisation" [Waddington(1942)]. This robustness to change within (e.g. mutations) and outside (e.g. environmental influences) usually involves some kind of self-regulation. Having said that, sometimes relatively similar genomes can account for large phenotypic differences, due to morphogenesis or development – for example segment number in closely related species of centipedes. The following section introduces some general developmental mechanisms that are thought to confer evolutionary advantages.

2.2   Evolution of Development

Evolutionary Developmental Biology, 'Evo-Devo' for short, studies the relationship between ontogenesis and phylogenesis, i.e. how developmental mechanisms determine the phenotypic variation arising from genetic variation, and their interconnection to evolution. Additionally any developmental model for a structure must be able to account for the development of earlier forms in the ancestors [Carroll et al(2001)Carroll, Grenier, and Weatherbee] – while at the same time evolutionary history constrains further evolution. Mechanisms believed to be crucial are developmental plasticity, modularity as well as heterochrony and heterotropy.

2.2.0.1   Developmental Plasticity

Under developmental plasticity, depending on environmental influences, a genotype can generate different phenotypes (or better: has a tendency towards one of several

---

[4] Prokaryotes are all unicellular, with few exceptions like myxobacteria, which have stages of multicellularity in their life-cycle.

Fig. 2.2 Modularity shows in *Drosophila* mutants: Small genetic differences lead to the development of modules out of place. Top: Normal adult fly. Middle: *Ultrabithorax (Ubx)* mutant - three mutations disrupt *Ubx* expression resulting in an extra pair of wings. Bottom: *Antennapedia* mutant - legs are grown in place of antenna. From DNA to Diversity by Sean B. Carroll, et al. (c) 2001 by Blackwell Science. Used with permission.

phenotypes, which can be expressed as a Gaussian distribution, the "norm of reaction"). That one genotype will often not produce exactly one phenotype was first realised in agriculture. The most illustrative example however is probably from social insects, where the development into e.g. worker and guard "morphs" depends on the treatment of the eggs by queen and workers. Another example is explorative over-production followed by positive or negative reinforcement (e.g. organization of the nervous system, but also within the cell in microtubule assembly) – This can add stability as small changes could be detrimental otherwise, and reinforcement can help accommodate to the environmental setting. Phenotypic or developmental plasticity can be of adaptive value and then, of course, has an evolutionary impact [West-Eberhard(2003)]. As mentioned above, evolved canalisation can restrict such variability, so that many genotypes in many environmental conditions consistently produce "one" phenotype.

2.2.0.2 Modularity

A module or compartment is defined as a (dynamic) structure with a boundary that is weakly coupled with its environment. Modularity allows for repetition, segmentation and symmetry. In biology, modularity is found everywhere: on the molecular, cellular, tissue, and organ organisation levels, and even in the life cycles of organisms (think larval and adult stages of insects). A phenomenon that recurs so often in nature must almost certainly provide an evolutionary advantage. The most popular explanation being that it allows for a somewhat independent evolution of the parts. Modularity is also directly connected to the creation of novelty via duplication followed by divergence. Duplications come at low cost, as redundancy is believed to be selectively neutral in many cases. Later on in evolution the duplicated parts are free to specialise on new functions as these are a lot less restricted by selection pressure now [Conrad(1990)]. A popular example are insect segments which are believed to have been duplicated earlier in evolutionary history. For an overview of prominent manipulation experiments showing modularity in the fruit-fly *Drosophila* see figure 2.2. Moreover, it is likely that differentiated multicellular organisms originated from colonies of unicellular Eukaryotes with almost identical behaviour [Buss(1987)]. Other important examples are more genetic and thus discussed later, in section 3.1.1.

2.2.0.3 Heterochrony and Heterotropy

Heterochrony refers to the change in the timing or rate of developmental events often observed between closely related species. For example head and brain growth stops shortly after birth in chimpanzees while it continues for several years after birth in humans. The related concept of heterotropy refers to a spatial instead of timing change in gene expression. So heterochrony and heterotropy presume modularity, as they describe changes in a somewhat independent module in one species compared to the development of the homologous (corresponding) module in a related species [West-Eberhard(2003)]. Again, these concepts are discussed in some more depth in later chapters.

2.3 Evolutionary Algorithms

Relatively early in the history of computation, engineers had the idea to harness the power of evolution for optimisation purposes [Fogel(1998)]. The simulation of evolutionary processes (heredity, variation, selection) proved particularly well-suited to complex, multidimensional problems too big to search exhaustively. As with all global search heuristics, evolutionary algorithms (EAs) are not guaranteed to solve all problems perfectly, but have fewer restrictions than most problem-solving algorithms. Several similar approaches were developed, most notably Genetic Algorithms (GAs) [Holland(1975)], Evolutionary Strategies [Rechenberg(1971)], Evolutionary Programming [Fogel et al(1966)Fogel, Owens, and Walsh] and Genetic Programming [Koza(1992)]. Here, the focus will be on GAs, however large parts of the analysis also apply to the other methods. This overview is based on [Knabe et al(2010)Knabe, Wegner, Nehaniv, and Schilstra], [Mitchell(1998)] and

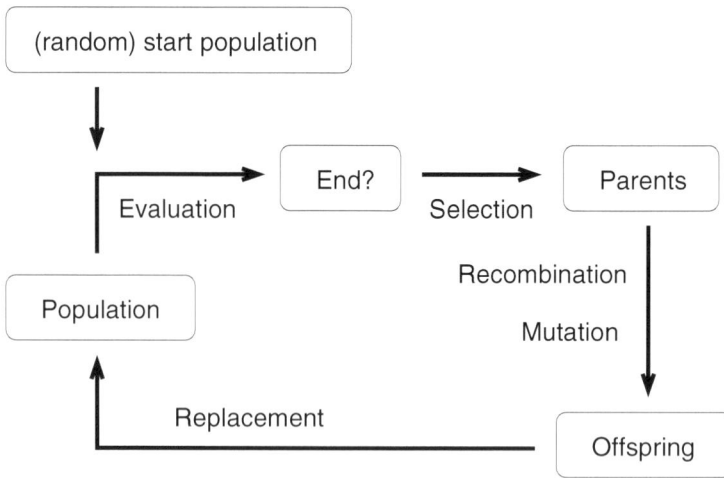

Fig. 2.3 Schema of the basic evolutionary algorithm cycle, which is - at this level - pretty close to the "real thing" (cf. fig. 2.1).

[Back et al(1999)Back, Fogel, and Michalewicz], where the interested reader can also find more variants and references.

The application of a GA to a specific problem requires tailored specification of

1. the genetic representation (an abstract representation of the solution domain that allows modification by stochastic variation operators, such as mutation and recombination),
2. a fitness function to evaluate candidate solutions, and
3. appropriate reproduction, selection, and termination strategies.

Once these have been specified (described below), a canonical GA does the following: The set of candidate solutions, referred to as population, is initialised. Usually this is done by randomly generating genomes, which in many cases are simply sequences of 1s and 0s. Next, the genomes are mapped onto the solution domain yielding the so-called phenotypes. These phenotypes are then evaluated. Depending on their performance as determined by the fitness function some are selected to be parents. The genotypes of the chosen individuals are recombined and mutated to create offspring. Their genomes replace others in the population, so that the composition of the population changes in part or completely. These steps, one generation, are iterated until a termination criterion is fulfilled (see fig. 2.3 for a schematic cycle).

GA design choices are of paramount importance for its performance, particularly the fitness function and genetic representation. The latter, as a central point of this work, will resurface time and again. As stated above, the minimum requirement for a genetic representation is to encode candidate solutions in a way that allows modification by "variation operators". Such operators, like mutation and recombination, only make sense when they add some amount of variation while preserving most traits – without heritability a random search would be just as good.

The issue of representation is directly connected to how variations in the genome are mapped to changes in phenotypic traits – the "genotype-phenotype map". For example, genomes can be sequences of real numbers or binary values, and the latter can be decoded to a number using base-2 or Gray encoding[5]. In this work it is argued that in this area GAs are often too simplistic, lacking expressiveness and non-linearity. This is explained in more detail further below, e.g. in section 2.4.

2.3.0.4   Fitness Function and Fitness Landscape

When applying a GA one usually has a clear problem that should be solved, so individual solutions are evaluated against this *a priori* target in a so-called fitness function. The fitness function is probably the most unrealistic part of a GA, as in nature only survival and reproduction (in an unpredictable ecosystem, shared with others) count – by comparison GAs are optimisation oriented, always searching for solutions closer to the target. A well defined fitness function allows for a visual representation of the kinship / fitness relationship search space[6]: So-called fitness landscapes are defined by the kinship of genotypes and fitness values. Kinship relates to the number of "variability steps" (where one variability step is usually equal to one point mutation) are required to change one genotype into another: the smaller the number of steps, the greater the kinship between the genotypes. In the example fitness landscape in fig. 2.4 all possible genotypes have been mapped onto the x/y plane, with genotypes the closer together the greater the kinship. Plotting the fitness for each genotype point as height (z axis) gives a landscape with peaks and valleys. This landscape metaphor can be useful[7] to get an idea of how the EA population "moves" about in this space and how a trade-off between exploration and exploitation can be found: One candidate solution is equivalent to a point on the landscape and its offspring another point, which can be close-by or far away. This generational change on the landscape can be interpreted as movement. A smooth hill will usually allow "climbing" uphill easily, while parameter settings determine the chance of bigger "jumps" (multiple mutations, crossover) and their size i.e. how likely the EA is to escape a local optimum. The landscape's shape is partly determined by the problem, but the mapping from genotype to phenotype and the fitness function play a very important role. For example it is easy to see that a Boolean fitness function (1 if problem solved, 0 otherwise) will result in a single peak and no gradient at all,

---

[5] In base-2 the bits represent increasing powers of two, with the rightmost bit encoding 2^0, so that flipping (or mutating) a bit on the far left can change the encoded value a lot. Gray or reflected binary encoding on the other hand ensures that only one bit has to be changed in order to represent the next integer, e.g. 00 binary maps to 0, 01 maps to 1, 11 maps to 2, 10 maps to 3 – The "distance" in genotype space corresponds to the distance in phenotype space (see below).

[6] The problem of well definedness in biological reality did not prevent evolutionary biologists from using the same process of thought, actually one of them, [Wright(1932)], came up with the idea.

[7] Mostly for conceptual thinking in toy problems, as plotting the full landscape is clearly impossible for complex problems.

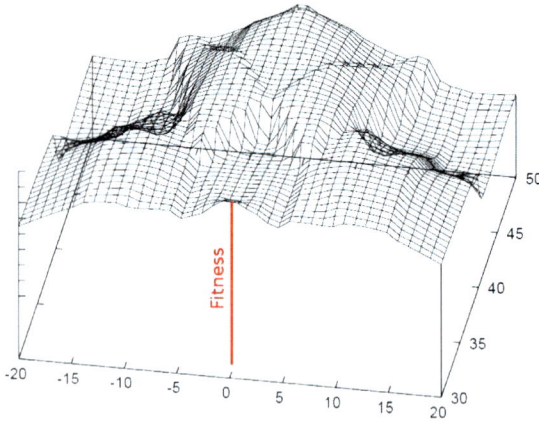

Fig. 2.4 Example fitness landscape

so an EA can become random search. Furthermore, low heritability of (phenotypic) traits will result in a rugged landscape. In nature fitness is definitely not static or known *a priori* as it depends on a whole ecosystem. However even for dynamic fitness functions (which are sometimes also used in EAs), the landscape metaphor can be useful as changes to the "terrain" might follow a particular pattern. Also, variable fitness can make a population more mobile in fitness space, leading to more independence from the randomly chosen start population. How the population moves about in this space is determined by selection and recombination strategies as well as parameter choices.

2.3.0.5   Selection

Natural selection favours fitter individuals for reproduction, but less fit individuals still have a (finite) chance of reproducing – thinking of a fitness landscape with several hills (multiple maxima separated by unfavourable intermediates) it is obvious that without stochastic selection the population could not "move downhill" and thus reach new hills, the chance of getting stuck in local optima is much higher. To use this principle for GAs, despite of the restrictions of not having an ecosystem but isolated fitness values, several methods have been developed, of which the most widely used ones will be discussed. Roulette wheel or fitness proportionate selection has a stochastic element – every individual gets a region on the wheel assigned: the better the fitness, the bigger the region. The roulette is spun and the roulette-wheel pointer selects the solution when the roulette stops, see figure 2.5. Since fitter solutions are represented by larger sectors, their chance of being selected is higher. Formally, if f_i is the fitness of individual i in the population of size N, its region (and accordingly reproduction probability) is $r_i = f_i/(\sum_{j=1}^{N} f_j)$. Here the reproduction probability is relative to the fitness of all other individuals in the population, unlike in tournament selection. For tournament selection a number of solutions is randomly picked from

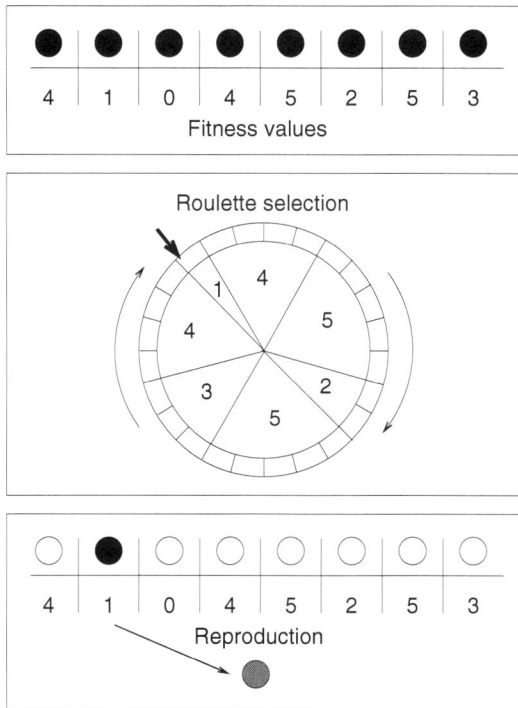

Fig. 2.5 The roulette selection mechanism with probability in proportion to fitness: Every individual gets a region on the wheel assigned; the better the fitness, the bigger the region. The roulette is spun and the roulette-wheel pointer selects the solution when the roulette stops.

the current population and from these the fittest solution(s) are chosen for reproduction, cf. fig. 2.6. Finally there is elitism. Based on the fitness value, a number of solutions are kept unchanged for the next population, while the remainder of the population is replaced with offspring, for example generated with one of the methods described above. Elitism is very unrealistic, but has the advantage that the best solutions cannot be lost by chance and that the best fitness value will never decrease over generations.

2.3.0.6 Variation Operators

To generate offspring that resembles the parents with some random variation, recombination (also called crossover) and mutation are employed. For binary genomes, mutation is simply flipping a bit, while for real valued genomes the original number is usually modified by a random value drawn from a Gaussian distribution. Mutations can be complemented by crossing over the genomes of two (or more) parents or by non-standard operators like duplication, inversion and deletion – see fig. 2.7 for

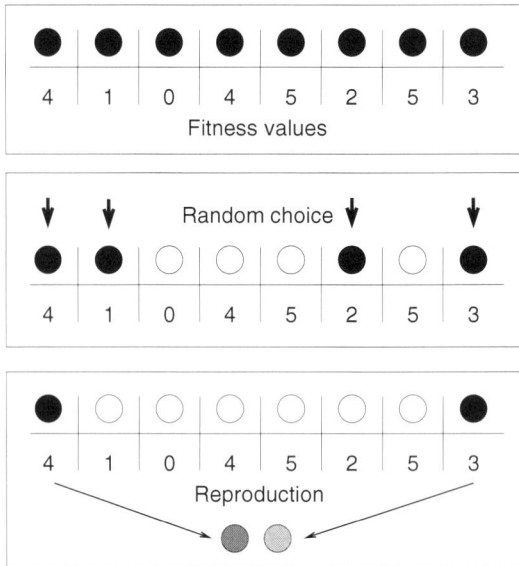

Fig. 2.6 The tournament selection mechanism: From the current population a number of individuals is randomly chosen. Of this group the fittest get to reproduce.

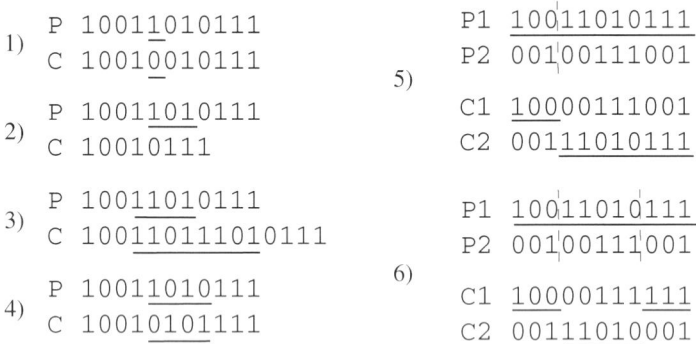

1)
P 10011010111
C 10010010111

2)
P 10011010111
C 10010111

3)
P 10011010111
C 100110111010111

4)
P 10011010111
C 10010101111

5)
P1 10011010111
P2 00100111001
C1 10000111001
C2 00111010111

6)
P1 10011010111
P2 00100111001
C1 10000111111
C2 00111010001

Fig. 2.7 Overview of variation operators; examples with binary genomes: 1) point mutation, 2) deletion, 3) duplication, 4) inversion, 5) single point crossover, 6) two point crossover. P (P1, P2) is for parent (1, 2) and C (C1, C2) for child (1, 2).

examples. The crossover type depends on the number of crossover points: 1-point to n-point crossover. The points can be chosen randomly or at "preferred points" in the string to e.g. implement modularity. Crossover can be disruptive to fitness if very different genomes are combined, as usually information on the genome is not totally independent. However on nearly converged populations, with relatively similar genomes, the effect can be more advantageous [Harvey(1992)]. Moreover, when genomes of different length are allowed, equal or unequal crossovers are possible where the crossover is either at the same position in the parents or at different positions. Duplications can then be realised by randomly choosing part of a sequence which is copied and inserted elsewhere. Similarly, in deletion a region is randomly chosen and removed from the genome. Generally selection and differential reproduction decrease diversity in the population, while mutation and recombination increase diversity. Recombination can be particularly useful in "pulling" the population away from local optima.

Finally the offspring replaces all or at least a large part of the population, as in almost all GAs population size is kept constant, attempting to simulate scarce resources and death. Offspring might replace their direct parent(s) or simply the whole population at once, but some more sophisticated schemes have been devised. To avoid "crowds" (many similar individuals) getting stuck in local optima, [de Jong(1975)] invented a "crowding" operator. This operator lets new offspring replace only the genetically most similar individuals in the population, thereby helping to ensure diversity. Furthermore, "fitness sharing" [Goldberg and Richardson(1987)] could prevent a loss of diversity: The population is split into neighbourhoods depending on the similarity of the individuals. Fitness values decrease with the number of individuals in a neighbourhood, thereby rewarding greater diversity. A comparison of these and similar methods can be found in [Mahfoud(1995)].

Often it is worth tuning the parameters (mutation and crossover rate, population size, etc.) to achieve a balance between exploration and exploitation suited to the problem at hand. However, there are no general laws to calculate the perfect settings and the parameters are not independent so they cannot be optimised one by one. Techniques have been proposed to regulate parameters during an evolutionary simulation, the most widely used of which is simulated annealing (SA), also an optimisation method in its own right. In SA a temperature value is introduced which is coupled (usually in an exponential fashion) to the variability parameters. Over evolutionary time the temperature and accordingly variability rates are reduced. The idea behind this is to explore wide areas of the search space initially while focusing on the local improvement of discovered solutions later on.

The following section discusses evolvability, which strongly depends on the GA design choices.

2.4 Evolvability

Evolutionary change depends on what is variationally possible, probable, and permissible [Carroll(2008)]. While the latter is mostly governed by selection, evolvability is mostly concerned with the former two: it describes a system's ability to evolve in an appropriate but unknown direction [Kirschner and Gerhart(1998), Nehaniv(2003)]. Biologists usually study the end-products of evolution, after apparently favourable variations have occurred. Engineers and evolutionary biologists face the problem of how these variations were possible.

While there is selection for adaptive variants there is no direct selection for the continued generation of adaptive variation – evolvability is a by-product of evolutionary processes and their influence on the genotype-phenotype map and genetic representation. The genotype-phenotype map describes how a genome is decoded into an individual. Thus, it influences also how genotypic variation is mapped to phenotypic variation [Wagner and Altenberg(1996)].

[Nehaniv(2005)] offers a formal definition of evolvability as "the capacity of a population to generate adaptive heritable genotypic and phenotypic variation". In this definition, "adaptive" is understood as fitter than any currently existing.

In evolutionary algorithms the genotype-phenotype map is often a very explicit, fixed linear function: take for example binary encoding, directly mapping onto phenotypic properties, e.g. say bits 0-7 of the genome are the binary notation of body length. In nature, "mapping" – differentiation, growth and morphogenesis – is a highly dynamic process, open to environmental influences. Evolution is always opportunistic, "using what it can get a hold of, tinkering with existing parts", therefore it needs complexity to allow for redundancy, compartmentalisation, neutrality. For example it is assumed that bird feathers evolved as a means of insulation or ornamentation and later on, initially as a by product, where used for flight [Prum and Brush(2002)], so there was a shift in function of a character. (This shift in function of a trait during evolution is usually referred to as exaptation.) In computer models the complexity needed for such twists is rarely present. Instead, intermediate steps between good phenotypes are usually not rewarded, which results in isolated "hills" on the fitness landscape. So heritability of traits is important for locally smooth evolutionary dynamics: small changes to the genotype should often result in small changes to the phenotype (and similar performance). Neutrality, changes to the genotype without effect on phenotype (performance), can make evolutionary search less sensitive to initial conditions, and thereby convey evolvability advantages, as [Huynen et al(1996)Huynen, Stadler, and Fontana] have shown. As mentioned above (section 2.2.0.2), modularity allows for independent optimisation of traits and thus increases evolvability, but where does it come from? Increased modularity is not directly selected, but could result from opposed selection pressures. Imagine a situation in which the expression of two traits is jointly controlled, and one comes under selection pressure to change while the other has greater fitness when left unchanged. Then it is easy to see that an uncoupling of the traits would be favourable [Wagner(1996)]. If one assumes that duplication is initially nearly neutral to fitness but frees the original of strong selection pressure so that it can diverge towards new functionality as for example [Calabretta et al(2000)Calabretta, Nolfi, Parisi, and Wagner] do, then duplication and divergence may also give rise to increased modularity. However it appears more likely that out of the many duplication events that occur, few are truly neutral. Only those duplications that disturb the organisation least (presumably modular ones) persist [Altenberg(1995)].

Chapter 3
Genetic Regulatory Networks

With the discovery of DNA a full understanding of the program controlling all cells[1] seemed in reach, as most researchers assumed that an organism is a direct reflection of its constituent genes[2].

A gene is the basic functional unit of heredity, however often does not encode information in a direct way: Gene activation at one point can regulate expression levels of genes at other locations or even feed back to itself, and in multicellular organisms cell interactions add another level. So the genome holds compact generative information. Accordingly, knowing the sequence of the bases in an organism's DNA is not enough – in order to understand how organisms are built and sustained we have to understand the dynamics of gene expression and regulation. However, *in vivo* analysis of these complex systems still poses many problems. Modelling and simulations can help understand as well as discover regulatory principles. Evolving artificial gene networks *de novo* can give new insights into their computational potential and the constraints that their real counterparts are subject to.

First, current ideas about the origin, structure, and dynamic behaviour of natural GRNs are discussed, followed by a review of how GRNs are modelled and simulated. After this overview, the last section of this chapter introduces the particular GRN model that was used for all experiments reported in this work.

3.1 Natural Genetic Regulatory Networks

In living cells, genes are contained in linear stretches of DNA that are organised in chromosomes. An organism's genome is the set of all of its genes or, more precisely, all DNA in a cell. Generally speaking, a gene holds two distinct pieces of

---

[1] There are a few exceptions, for example *erythrocytes*, red blood cells, in mammals lack DNA in their mature state.

[2] This simplistic view can still be found in many yellow press articles that present "a gene for" some complex property or even behaviour, in part due to a mix-up of the informational and bio-physical meaning of 'gene' as well as some of their reader's preference for simple explanations.

J.F. Knabe: Computational GRNs: Evolvable, Self-organizing Systems, SCI 428, pp. 19–43.
springerlink.com   © Springer-Verlag Berlin Heidelberg 2013

information: the "code" for its gene product (or products), and the "combination lock" to its expression. A gene product (GP) is a macro-molecule that has been constructed on the basis of the information present in the coding region of the gene, and may be a polypeptide (protein) or a polyribonucleotide (RNA). A gene is said to be expressed when its products are being constructed. Within an individual organism, expression levels - quantities of GP present at a particular time - may vary over time, and from cell to cell. This variation occurs in spite of the fact that all cells within the organism have the same genome, and therefore contain the same information. In the context of gene regulation, gene products are usually divided into two classes: structural and regulatory. Structural GPs contribute directly or indirectly to the structure of a cell and its functioning, for example as enzymes in the breakdown of a food source. Some structural GPs, such as certain participants in basic cellular metabolism, are continuously present in all cells; these are often referred to as housekeeping genes. Most genes, however, are only expressed to a significant level in specific cells at specific times: their expression is normally repressed, and only switched on under certain conditions. The differential expression of these so-called conditional or facultative genes determines the differences between cell and tissue types. Conditional gene expression is at the basis of the development of an embryo into a fully differentiated multicellular organism, and also plays a central role in cellular adaptation and response [Davidson(2001), Davidson(2006)]. This was first fully realised by [Jacob and Monod(1961), Jacob and Monod(1963)] when they discovered within the bacterium *E. coli* a gene that switched between metabolic modes depending on presence of lactose in the environment, see fig 3.1. Regulatory GPs are involved in the repression or activation of gene expression, and are sometimes called transcription or trans-regulatory factors (TFs). The expression of a gene may be controlled by the products of other genes, and also by its own GP, and one GP may be involved in the regulation of multiple genes. Regulation is mediated through binding of TFs to so-called cis-regulatory sites on the gene whose expression they control. Physically, a cis-regulatory site is a stretch of DNA that has a high affinity for a specific TF or TF complex. These sites are usually located relatively closely to the coding area. If TF molecules are present in a given cell over a certain period of time, they will occupy their cis-regulatory sites for at least part of that time. Control is achieved when the occupancy of a cis-regulatory site by a TF locks ("represses") or unlocks ("activates") the construction process of new molecules of the gene's product. Generally a high affinity will slow down dissociation, leading to a long occupancy.

GPs are produced in a series of highly complex consecutive processes that start with transcription of the information contained in the coding area of the gene into an RNA transcript. Transcription is initiated by RNA polymerase, an enzyme, binding to specific DNA sequences known as promoters near coding areas. RNA polymerase will preferentially read and transcribe DNA until another specific DNA sequence, known as terminators, at the end of gene is encountered – when the transcript is released. RNA that is later translated to give rise to protein is called messenger RNA (mRNA). Other RNA is not translated and has housekeeping or regulatory functions. In Eukaryotes mRNA undergoes complex post-transcriptional modification

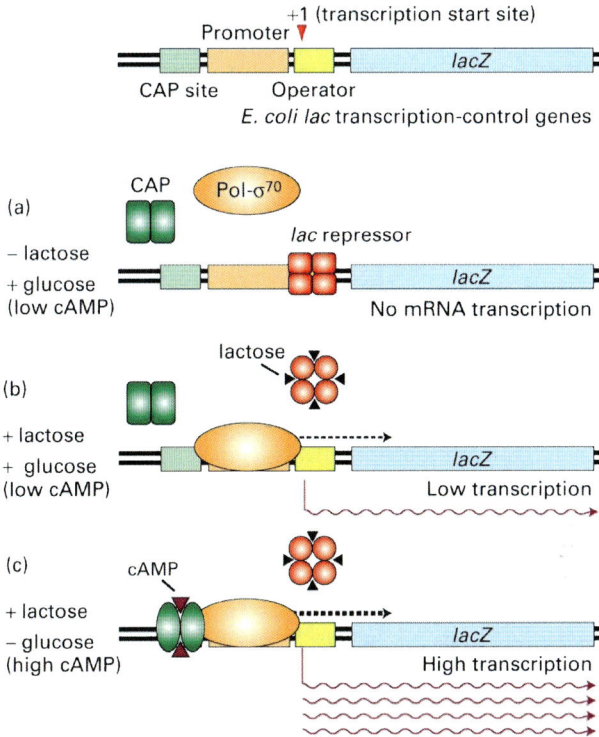

Fig. 3.1 *lacZ* transcription control in *E. coli*: *E. coli* need glucose as energy source. They can produce an enzyme (β-galactosidase) to digest lactose into glucose. However, it would be inefficient to produce enzymes when there is no lactose or glucose is readily available. The transcription of *lacZ* gene, which encodes for β-galactosidase, is regulated by lactose and glucose levels: (a) *E. coli* express a protein called the *lac* repressor. In the absence of lactose, this protein can bind to DNA near the promoter ("operator") for the *lacZ* gene. This prevents RNA polymerase (Pol-σ^{70}) from binding, and no transcription is possible. CAP (Catabolite activator protein) does not bind to the corresponding site as it is in an inactive state. (b) If lactose is present, the lactose binds to the lac repressor, inducing a conformational change that causes the repressor to fall off of the DNA and transcription can proceed at a fairly low level. (c) Low levels of glucose in a cell lead to the formation of cAMP (cyclic adenosine monophosphate). cAMP can bind to CAP, causing a conformational change allowing it to bind to the CAP site, where it accelerates RNA polymerase binding. This leads to increased transcription of the *lacZ* gene.
From *Molecular Cell Biology*, 5/e by Harvey Lodish, et al. (c) 1986, 1990, 1995, 2000, 2004 by W. H. Freeman and Company. Used with permission.

or maturation, so the earlier form is referred to as precursor mRNA or pre-mRNA. Eukaryotic mRNA maturation involves several processes, such as capping and polyadenylation which are mostly related to mRNA stability. In a process called splicing, regions termed "introns" of the pre-mRNA are lost, while "exon" regions are retained and joined together. "Alternative splicing", the controlled selective excision of introns from one pre-mRNA may yield several mature mRNA "isoforms", each with a somewhat different nucleotide sequence [Sammeth et al(2008)Sammeth, Foissac, and Guigo]. It has been speculated that alternative splicing confers evolutionary advantages, firstly because it allows for rapid generation of proteins with novel structure and function, and secondly because the regulation of specific splicing patterns may be conserved.

This is followed by transport of the mature product: Translation of the information in the mRNA into a polypeptide chain, followed by post-translational modification into a mature, albeit not necessarily immediately active, protein. Activation (and subsequent inactivation) of proteins may require further modification steps or complex formation[3]. Extensive quality control, and reliable repair of faults in DNA replication ensure the persistence of the information content of genes over many generations. In contrast, RNA and protein molecules have limited lifetimes. Not only are they, in the absence of major repair mechanisms, subject to thermodynamic decay (loss of functionality through spontaneous breakage of covalent bonds); they are often actively being broken down (by other gene products that catalyse the breakdown). In general, a capacity for rapid breakdown contributes to a cell's capability to respond rapidly to incoming signals, and to adapt efficiently to changed circumstances.

Like gene transcription, the processes involved in mRNA translation, RNA and protein maturation, activation, inactivation, and breakdown are all highly regulated. In the following, however, the focus will predominantly be on the regulation of transcription, not only because it forms the basis for all higher-level regulation, but also because it can be highly adaptive, and has the ability to integrate the information contained in a large amount of incoming data.

A cis-regulatory region (a collection of cis-regulatory sites for the same gene) may contain binding sites for many different TF types, each of which will affect the rate of transcription differently. Importantly, like structural GPs, TFs may act "synergistically": their combined effect can be very different from the sum of their individual effects - an example is shown in fig. 3.2. A set of cis-regulatory binding sites for TFs that act in some kind of synergetic combination is sometimes called a cis-regulatory module. A simple example of synergy is a situation in which only the physical complex of two different TFs (a TF hetero-dimer) is able to repress transcription, whereas either of the constituents of the complex have no effect. In that case, repression will only occur under conditions in which both TFs - which, as the products of different genes, may require different sets of conditions for their own expression - are being expressed. The receptor TF might undergo conformational changes (changes to its three dimensional shape), which can affect its functional state or affinity. [Gerhart and Kirschner(1997)] noted that "Eukaryotic transcription factors often have limited affinity and sequence specificity on their own and require the presence of other factors to confer stability and specificity in DNA binding".

---

[3] A broad and more detailed introduction with all known chemical details can be found in textbooks on biochemistry or molecular biology, such as [Alberts et al(2002)Alberts, Johnson, Lewis, Raff, Roberts, and Walter].

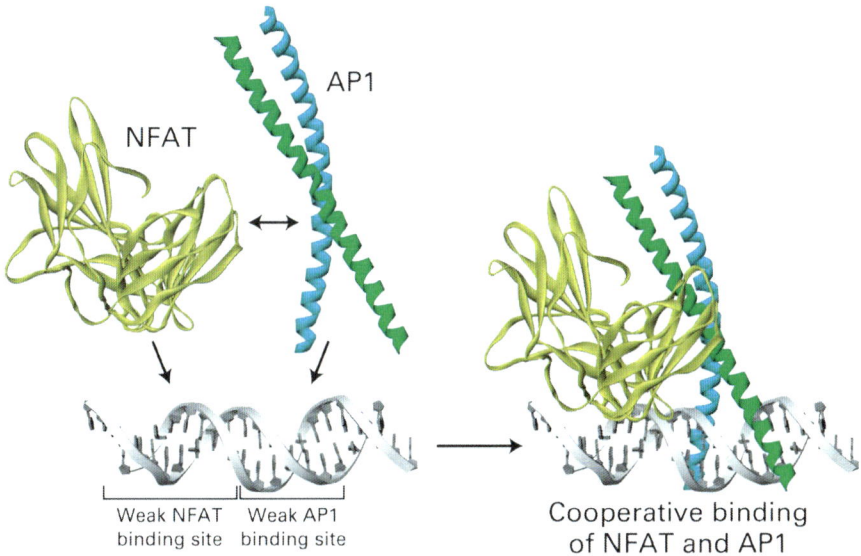

Fig. 3.2 Synergistic binding of Transcription Factors example. "Activator protein 1" (AP1) and "nuclear factor of activated T-cells" (NFAT) interaction results in a far stronger binding than would be possible for each of them on their own. From *Molecular Cell Biology*, 5/e by Harvey Lodish, et al. (c) 1986, 1990, 1995, 2000, 2004 by W. H. Freeman and Company. Used with permission.

3.1.1   Developmental-Genetic Toolkit

This section introduces the genetic basis (and evolution) for some of the developmental mechanisms described in section 2.2. Here the molecular basis of development is discussed, while spatial, cellular mechanisms are left for chapter 6.

The large scale effects of genetic "switching" and TF synergism are that regulation is often hierarchically organised. A relatively small number of "master control genes"[4] start cascading control sequences thereby choosing the program expressed by the cell (and possibly neighbour or daughter cells) – differentiation. This is especially important during the early stages of embryonic development. For example, [Halder et al(1995)Halder, Callaerts, and Gehring] found that eye growth in *Drosophila* is triggered by a single gene, called PAX6. Apart from the group of PAX genes, homeobox or HOX genes [McGinnis et al(1984)McGinnis, Levine, Hafen, Kuroiwa, and Gehring] are important in specifying the identity of particular body regions. These genes are highly conserved over evolution, as changes to them would often have detrimental effects: An individual failing to express PAX6 would not

---

[4] The term "master control gene" might be a bit misleading, as hierarchical is not meant to imply that there is a strict top-down command structure without feedback. Rather some regulatory events propagate through large parts of the network, changing dynamics consistently and in the long term.

Fertilized egg   **Early embryo**

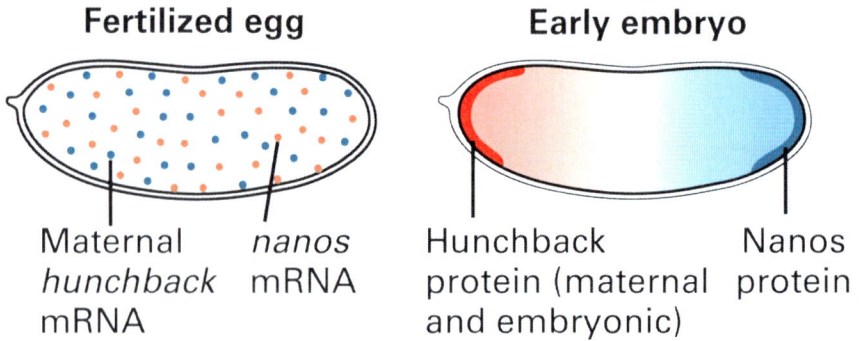

Maternal   *nanos*   Hunchback   Nanos
hunchback   mRNA   protein (maternal   protein
mRNA   and embryonic)

Fig. 3.3 *Drosophila* embryo gradient setup schema. From a relatively uniformly distribution of mRNA molecules in the fruit fly egg, during morphogenesis, two gradients are built up in the early embryo. From *Molecular Cell Biology*, 5/e by Harvey Lodish, et al. (c) 1986, 1990, 1995, 2000, 2004 by W. H. Freeman and Company. Used with permission.

develop an eye. Striking evidence of this evolutionary restriction is that PAX6 GP can be exchanged between species as distant as mouse and fruit fly and still trigger eye development. Such an hierarchical arrangement can also loosen evolutionary constraints: Genes and regulation hierarchically below a master GP may be independent of other development. Then, changes only affect one module, thus reducing the possibly devastating effects of pleiotropy (one gene influencing several traits).

Master control genes are often regulated by so-called morphogens[5]: GP gradients, established along embryonic axes (see fig. 3.3 for an example and section 6.1.2 for details.). Cells close to the source of morphogen will receive high levels of morphogen and will express both low- and high-threshold target genes. In contrast, cells far away from the source of morphogen receive low levels of morphogen and express only low-threshold target genes. The identity of particular body regions emerges as a consequence of the different combinations of target gene expression.

This aspect of developmental regulation also has evolutionary advantages: Changing the morphogen signal itself (spatially or its timing) or the regulatory thresholds can "move a module around". Such changes (in combination with other inter cellular signalling) are also thought to be the basis for heterotropy and heterochrony [Buss(1987)].

Modularity can be of evolutionary advantage when duplication leads to redundancy but allows later divergence (as described in sections 2.2.0.2 and 2.4). As mentioned before, duplication has to be selectively almost neutral[6] (or

---

[5] Regulation of control genes by morphogens is not necessarily direct. Rather it often involves binding of the morphogen to cell receptors and signal transduction communicating the morphogen level.

[6] It is not clear how often duplications are nearly neutral. For example not the full gene with all regulatory elements may be duplicated, as some of the regulating parts can be very far away from the coding region. Furthermore two copies may produce twice the amount of GP. Such differences will often make duplication not neutral.

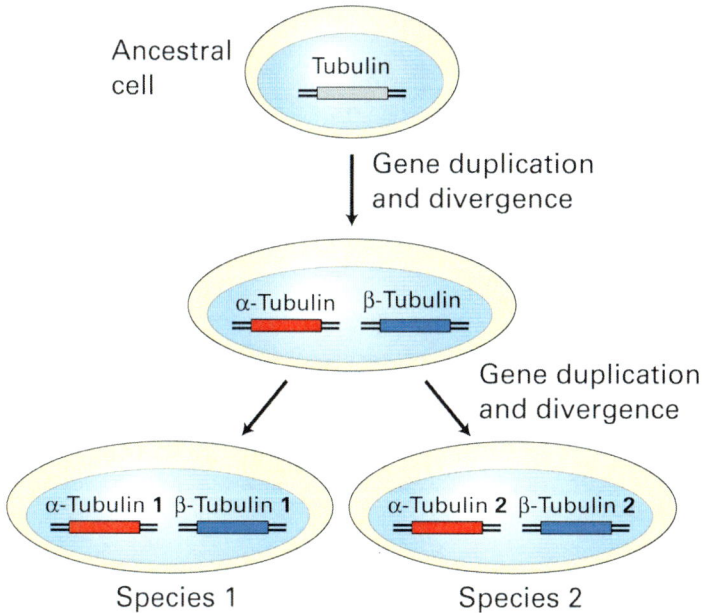

Fig. 3.4 Duplication and Divergence – Tubulin example. The genes after the first step are called paralogs, as they were separated by gene duplication. Genes separated by speciation, as is the case after the second step, are orthologs. From *Molecular Cell Biology*, 5/e by Harvey Lodish, et al. (c) 1986, 1990, 1995, 2000, 2004 by W. H. Freeman and Company. Used with permission.

advantageous) initially for the mechanism to work. This seems to have been the case for master control genes: In vertebrates (humans etc.) there are four clusters of homeobox genes (about ten genes each, on different chromosomes). These clusters have been formed by two whole genome duplications early in evolution. However these gene clusters are not identical nowadays as they have acquired differences over time. The importance of duplicating complete bits of genetic information for the evolution of biological complexity was already pointed out by [Ohno(1970)], see also [Maynard Smith and Szathmáry(1998)]. Ohno put emphasis on whole-genome duplications while it is now, with better techniques, becoming ever clearer that "both small- and large-scale duplication events have played major roles" [Taylor and Raes(2005), p. 320]. The history of homeobox genes is very well characterised, from the ancestor of Cnidaria (jellyfish etc.) with two of them in one cluster via many small and large-scale duplications to vertebrates with four duplicates (paralogues) of ten homeobox genes [Carroll et al(2001)Carroll, Grenier, and Weatherbee]. An example of duplication and divergence is shown in figure 3.4 (although tubulin is not a master control gene).

3.2   Genetic Regulatory Network Models

Owing to the interest in GRNs combined with the complexity of the systems and problems in conducting *in vivo* experiments, there has been a lot of work on finding good GRN models. As with all modelling, not every molecular detail can be included, but behaviours of the system considered essential should be captured – there is a trade-off between model simplicity and functional fidelity. Accordingly, the appropriate level of abstraction will depend on the intended purpose of the model [Bornholdt(2005)]. Purposes are many and thus types of models can be arranged along several dimensions, which is why the developments will not be presented in strictly chronological order. Two broad goals can be distinguished, firstly the aim of matching and possibly quantitatively predicting biological gene expression data; in such models as much detail as possible should be included. Often differential equations (coupled, possibly stochastic, ordinary differential equations, ODEs), describing the reaction kinetics of the constituent molecules are used in this case. Sometimes also parameter estimation methods for closely modelling natural GRNs, predicting unknown parameters (e.g. reaction rates) or network structures (e.g. interconnections) are used; see for example the "repressilator" [Drennan and Beer(2006)]. (Notably there are some attempts to use ODE models for automatic engineering purposes, e.g. [Fleischer and Barr(1993)] tried hand-coded ODEs plus differentiation via IF clauses in a developmental model for the construction of neural networks.) An extended overview of quantitative modelling and simulation methods for GRNs can be found in [de Jong(2002)] and the more recent [Schlitt and Brazma(2007)]. In the following the focus will be on mathematical models that qualitatively capture the most distinctive features of biological GRNs.

When the goal is to capture general properties of GRNs however, like their parallel distributed computational and adaptive power, more abstract models are appropriate. In the most simple case such a conceptual or artificial GRN is represented as a directed graph, with vertices (network nodes) representing genes and gene products (GPs), and edges or arcs regulatory influences between them. In more elaborate models, genes, GPs and their precursors (nucleotides, amino acids, raw transcripts, modified transcripts, etc.) may have individual representations. GRN models may also be classified on the basis of their update rules, which specify how the various processes contribute to changes in component levels (transcription, translation, modification, breakdown). Components can update synchronously, or in some (possibly non-deterministic) order. Time may be modelled in equally sized, discrete steps, or as a continuum, and GP levels and transcription factor (TF) efficacies can also be modelled discretely or continuously. Crucially the model has to define how fast and to what extent component values change in response to changed regulation. The relationship between the amount of TF and the GP production rate may be linear, hyperbolic, or sigmoid: in the last two cases a maximum rate (horizontal asymptote) must be defined. In order to prevent unlimited growth, models in which the amount of TF affects the GP production rate usually define an upper limit for each GP level, or specify GP breakdown processes in more detail. Synergy, or cooperativity, in effects of TFs may be taken into account or not.

As was discussed in section 2.3, for models that will be used in optimisation pro-cedures (for example using Evolutionary Algorithms) the genotypic representation and mapping to GRN phenotype the algorithm works on can be crucial. Needless to say that again approaches vary widely, between "no genetic representation level at all" to solutions where gene interpretation is grounded in sequence representa-tion, mimicking as much of the biological reality as possible. GRNs can be created randomly, based on or biased by available data (structural or dynamical expression data), and either of them can be a starting point for evolving particular properties (like connectivity or dynamic behaviour).

The following tries to give an overview of the various approaches, their purpose and contribution. This context should allow positioning of the extended `BioSys` (or xBioSys) model, described afterwards. Where these models have been applied to related tasks as the ones investigated in the following chapters the particulars are discussed in the "related work" section of the respective chapter.

3.2.0.1 Survey of Genetic Regulatory Network Modelling Approaches

The earliest and simplest GRN modelling is from [Kauffman(1969)], who inves-tigated random Boolean networks (RBNs). These networks only consist of nodes representing genes (and their GPs, these are not represented separately) plus direc-tional connections between nodes representing regulatory influences. Every node can only be either on or off and is connected to another node or not. This means a gene is active or inactive (i.e. a gene product is present or not present) and a reg-ulatory influence exists or it does not, without degrees of activation or regulatory strength. Updates occur in discrete, synchronous, time steps: The state of each node at time $t + 1$ is computed as a Boolean function of the values at time t from nodes connected to the node via incoming arcs. A simple example is shown in figure 3.5. After a number of update steps the finite gene expression pattern necessarily has to repeat itself, and the RBN has then settled into a (possibly cyclic) attractor (see the digression on dynamical systems in section 4.1). Depending on the initial activa-tion state a given RBN can settle into different attractors (given that the RBN has several), which it will stay in unless external perturbations - changing at least one gene's state deviating from the normal update rules - move it to another attractor. Kauffman argues that cyclic attractors can be interpreted as corresponding to cell types or at least alternative, stable, GRN behaviour modes.

As he was interested in general properties of Boolean networks, e.g. how many attractors ("cell types") one can expect with a particular number of genes and de-gree of connectivity, initial investigations were undertaken with randomly connected networks, the so-called NK networks: Specified was only the number N of gene nodes, and the number K of inputs to each node (connections from other nodes). Which nodes served as input, the connectivity, was randomly chosen, and so were the Boolean functions of the K inputs that would determine the next state of a node. For Ks of two and three [Kauffman(1969)] found that RBNs: 1) On average fall into cyclic attractors whose length predicts cell replication time (as a function of gene number or N); 2) exhibit a number of attractors corresponding to the number of cell

$$
\begin{array}{cc|c}
3 & 2 & 1 \\
\hline
1 & 1 & 1 \\
1 & 0 & 1 \\
0 & 1 & 1 \\
0 & 0 & 0
\end{array}
\qquad
\begin{array}{cc|c}
1 & 3 & 2 \\
\hline
1 & 1 & 1 \\
1 & 0 & 0 \\
0 & 1 & 0 \\
0 & 0 & 1
\end{array}
$$

$$
\begin{array}{cc|c}
3 & 2 & 3 \\
\hline
1 & 1 & 1 \\
1 & 0 & 0 \\
0 & 1 & 1 \\
0 & 0 & 1
\end{array}
$$

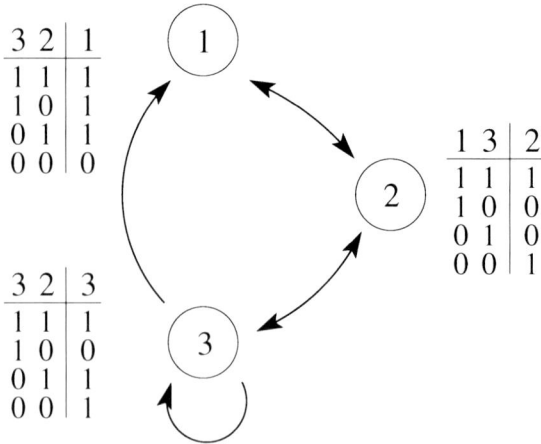

Fig. 3.5 Example of a Boolean Network with three nodes having two inputs each; the node's update functions are shown as truth tables: The activation value or state of a node in the next step depends on the current states of its input nodes. For node 1 this is (the states of) $3\ AND$ 2, for node 2 it is $1\ \neg XOR\ 3$ and for node 3 it is $3 \rightarrow 2$.

types in organisms (as a function of gene number or N). Furthermore, small external perturbations (switching relatively few nodes) to an RBN's state, if moving it to a different attractor at all, would only be able to move the RBN to some "neighbouring" attractors. This resembles the fact that (non-stem) cells can, depending on their own type, usually only give rise to a particular subset of cell types.

After these early successes Boolean networks were used for a lot of research projects owing to their simplicity, allowing simulation of RBNs with thousands of gene nodes and connections even on early PCs. A comprehensive overview is given in [Kauffman(1993)]. Notably, Boolean networks were later used with EAs to evolve a particular or a high number of attractors resp. cell types, [Kauffman and Smith(1986)] seems to be the first work of this kind. Owing to, again, the simplicity of the model, the fitness landscapes (see section 2.3) for a wide parameter range could be explored – one main finding was that RBNs with low degrees of connectivity (especially $K = 2$) are more evolvable. For large Ks landscapes were very rugged and thus EAs prone to get stuck in local optima. [Iguchi et al(2005)Iguchi, Kinoshita, and Yamada] have characterised differences in the ruggedness of fitness landscapes between random and scale-free[7] Boolean networks.

Later variants of the NK RBN formalism include using different input degree K for every node, as natural GRNs exhibit a scale-free distribution of input degree, where most nodes have few input connections, and few "hub" nodes have many. Another interesting variant restricts the Boolean functions to so-called canalising functions. In Kauffman's definition this means that at least one input value must

---

[7] In scale-free networks the probability $P(k)$ that a node connects with k other nodes follows a power law: $P(k) \sim k^{-\gamma}$, with γ between 2 and 3.

be able to control the outcome regardless of the other inputs. This restriction does not only reduce the size of the search space for EAs but also increases stability of the RBNs [Kauffman et al(2004)Kauffman, Peterson, Samuelsson, and Troein]. More recently Boolean networks have been used to replicate the dynamics of the *Drosophila* segment polarity network [Albert and Othmer(2003)][8]. In this case the Boolean network was of course not random, but (manually) designed to produce a particular expression pattern. Later [Chaves et al(2005)Chaves, Albert, and Sontag] found that the observed expression pattern changed dramatically when nodes were updated asynchronously (nodes would no longer update their states simultaneously but one after the other in a random order). However, with semi-asynchronous updates the original expression pattern could be recovered. The semi-asynchronous update schema is based on the idea that state changes do not all occur deterministically at the same time, but time-scales are preserved in nature; that is some processes can (almost) always be expected to finish before some other event occurs. For an analysis of update regimes see also [Gershenson(2002)], who showed that RBNs are more different depending on whether updates occur deterministically or not rather than depending on their synchronicity or asynchronicity. Boolean networks have been very successful as generic GRN models, e.g. to analyse dynamical systems features like the identification of basins of attraction, but one has to keep in mind that they are highly abstract.

Early bids for modelling continuous gene activation used connectionist "recurrent neural network"-inspired formalisms [Mjolsness et al(1991)Mjolsness, Sharp, and Reinitz, Vohradsky(2001)]. These however were still lacking a straightforward DNA-like representation. For example, using such an approach, [Geard and Wiles(2005)] have the aim to evolve cell arrays that exhibit a static differential gene expression pattern – an approach that is inspired by the spatial protein expression patterns observed in early embryos, where these proteins often control cascades of other genes thus directing the subsequent functioning of the cell. Selecting for particular switching patterns the EA operates directly on the network structure, e.g. rewiring of connections occurs with a certain probability.

Still with interest in general properties of GRNs as dynamical systems, a sequence based model with template matching has been used by [Reil(1999)]. In this model genome representation as a sequence of digits plays a fundamental role: The number of genes and their inputs is not fixed but a result of representation; for example '0101' starts a gene, meaning that the following six digits determine the GP. A GP is simply the six gene digits, increased by one each[9]. The sequence is then scanned for sections matching the GP template. Regulation occurs in an amount-independent way, so one instance of a GP is enough to turn on or off a gene – the paper states that activatory or inhibitory influence is determined, for example, by defining all genes ending with '1' as inhibitors. The resulting networks showed

---

[8] Actually the network was built to replicate the results from another, differential equation, model of the *Drosophila* segment polarity network.

[9] The paper does not mention what would happen to digit nine, supposedly the randomly created genomes did not include any nines

stable and cyclic activation patterns. While the dynamics are still very similar to
RBNs, sequence representation plays a crucial role: The number of genes and binding sites depends on the likelihood of the occurrence of '0101', which of course
directly correlates with sequence length.

[Banzhaf(2003)] investigated a related but more sophisticated model. He investigated bit-sequence genomes where genes are started by the pattern '01010101',
appearing with probability $2^{-8} \approx 0.0039 = 0.39\%$. Genes have a fixed length of
160 bits[10]; the 64 bit after the start section are interpreted as two binding sites, one
activatory, one inhibitory, of 32 bit each. The remaining 96 bits are used to determine
the 32 bit GP: The 96 bits are interpreted as three 32 bit sequences and the "majority vote" for each bit used as GP – a 3*8 bit example: '01011101', '01011101',
'10100010' would result in the GP '01011101'. With 32 bit per GP, chances for an
exact match of GP and regulatory binding site are slim of course, but perfect template matching is not necessary here. Affinity of a GP for a binding site is modelled
as (exponentially) stronger influence the more bits achieve an XOR (complemen-
tarity) match. The match u_j between GP and binding site is the number of bits set
in an XOR operation ($\bar{u}_{max}$ is the maximum match achievable, i.e. the length of the
bit patterns). With amount c_j of GP j and a scaling factor β the activatory (a_i) /
inhibitory (b_i) signal for GP i is computed as:

$$a_i = \frac{1}{N} \sum_j c_j e^{\beta(u_j^+ - \bar{u}_{max}^+)} \tag{3.1}$$

$$b_i = \frac{1}{N} \sum_j c_j e^{\beta(u_j^- - \bar{u}_{max}^-)} \tag{3.2}$$

Rather than a Boolean on or off, genes have continuous levels of activation. Given
the two inputs and constant Φ (not further specified in the paper), GP i changes as:

$$\frac{dc_i}{dt} = \delta(a_i - b_i)c_i - \Phi \tag{3.3}$$

GPs form a proportion of the overall GP amount, hence a flow term is used to as-
sure $\sum_i c_i = 1$. In this model a GP can be activating and inhibiting at the same time,
however no synergistic interactions are possible between GPs. Importantly regulation strength varies with the degree of match between GP and sequence, mimicking
affinity of GPs to binding sites. So in network terms a fully connected GRN results,
but with many very weak interactions. The dynamics (change of GP amounts over
time) of networks based on random sequences were analysed. Particular dynamic patterns could be evolved successfully. The possible benefit of a smooth, affinity based,
matching mechanism for evolutionary processes, namely that small changes to the
genotype are more likely to have small effects on the phenotype, is pointed out.

In a somewhat different line of work, evolvability and development as an indirect
mapping from compact genotype to complex phenotype are central. The problem

---

[10] Presumably start sequences within a gene's 160 bits are not interpreted as such.

of producing scalable evolutionary approaches for the automatic design of robot shapes and especially control systems resulted in EA researchers looking for inspiration from biology once more [Kitano(1990)]. A major problem they identified was that when the genotype directly encodes a neural network the number of coding bits increases exponentially with the number of nodes, repeated structures can not be represented in a compact way, and variation operators had a high chance of being disruptive. The influence of space and time on morphogenesis are neglected. A huge body of research with models that included some kind of developmental phase (which was usually highly abstracted, for example grammar or rule based), resulted. These are reviewed in [Stanley and Miikkulainen(2003)], while a review of early neural network encoding schemes can be found in [Nolfi and Parisi(1995)]. In these schemes, "genes" still directly encode particular parameters, e.g. the Cartesian position of a neuron and the branching angle of axons. Here, the focus will be on approaches that include GRNs explicitly, allowing such parameters to emerge from network dynamics.

[Jakobi(1995)] model is inspired by the growth of animal nervous systems. He evolved "developmental" neural networks for robot control. The GRN is sequence based (here: genomes of length 5000, four characters [a,b,c,d]), with genes of fixed length, where a gene start is indicated by 'aaaba' (appearing with probability $4^{-5} \approx 0.001 = 0.1\%$). This is followed by three characters which are mapped to the threshold stimulation needed to switch the gene on. The next three characters, the so-called link template defines which GPs regulate the gene. This is followed by another three characters, defining the structural class to which the GP belongs. Finally the 50 character GP is specified. Every gene produces one GP that belongs to one "structural" class: signals, movers, dendritics, splitters, differentiators, and threshold GPs. In addition, every GP may function as a regulator of other genes. The model strongly relies on template matching, take for example regulatory influence: To determine the influence one gene has on another, the former's GP is interpreted as a circular arrangement of characters. This is then rotated along the latter's link template until (if at all) a match threshold is surpassed, in which case a link between the two genes is created and the three characters diametrically opposite the matching characters are matched against another "arbitrary but fixed" template to determine the link's weight. Although the paper contains interesting ideas, it is, unfortunately, short on computational detail. External "protein sources" allow for orientation in a two dimensional space with growth starting from the left, where cells that later become input neurons are located, a middle space with inter-neurons and to the right future motor neurons. GP signalling however is immediate, i.e. a signal is weaker the further away from a source a receiver is, but there is no delay due to diffusion time. After the developmental phase the cells are interpreted as neurons in a neural network, with weights given by the level of weight GP and thresholds by the level of threshold GP in the corresponding cell. The neural networks evolved in this way could operate a small robot to avoid obstacles and similar basic behaviours.

[Dellaert and Beer(1996)] try a similarly complex GRN model but come back to RBNs, concluding that the search space in their original approach was too large for an evolutionary algorithm to find good solutions in reasonable time. Evolvability decreased, bringing back to mind that a developmental stage does not *per se* guarantee evolvable models.

[Eggenberger-Hotz(1997)] evolves (relatively regular) multi-cellular differentiated 3D shapes using a sequence based system. Genomes are sequences consisting of integers $\{0,1,2,3,4,5,6\}$. Such a sequence is divided into units of equal length. The function of each unit is determined by sequence and the last digit of such a unit, called marker. All units up to the first one with a '5' marker are considered regulatory binding sites, while all following units are structural genes until one with a '6' marker is encountered, after which regulatory sites follow up the next '5' and so on. Here the notion of a (structural) gene does not include regulatory sites, but defines them as separate entities. The block of regulatory units directly before a block of structural genes regulates the activity of the latter, so several binding sites may regulate the activity of several genes. In every step, the activity of the jth regulatory unit r_j of a structural gene is calculated as:

$$r_j = \sum_{i=1}^{n} aff_i \times conc_i$$

This is the sum over the affinity (aff_i) multiplied with amounts $conc_i$ of all GPs. The affinity aff_i of the ith GP with the jth regulatory unit is calculated as the difference of the integer representations of the GP's and regulatory unit's integer sequence (which can be a positive or negative number). So every GP can have an influence on activation in this model. The sum of these activations of one regulatory block is then used as input to a sigmoidal function:

$$a_k = \frac{1}{1 + \exp(-\sum_{j=1} r_j)} \tag{3.4}$$

Depending on threshold values, the following structural gene block is producing GPs g_k as:

$$g_k = \begin{cases} -1.0: & a_k < 0.2 \\ 1.0: & a_k > 0.8 \\ 0.0: & \text{otherwise} \end{cases} \tag{3.5}$$

Apart from being regulatory, GP amounts can also influence cell adhesion, communication, diffusion, cell division, death and 'cell searching'. GP sources placed in the three dimensional space allow for orientation of the cells.

The growth process however was not constrained by physical laws. In an extension of this work [Eggenberger(2003)] added elastic and viscous forces, showing that such physical rules reduce evolutionary complexity when compared to an approach where all details of the shaping have to be stored in the genome. Even earlier, [Hogeweg(2000)] used physical constraints, especially energy minimisation, in a multi-cellular developmental approach with RBN controllers. [Miller(2003)] also evolves differentiated patterns. His work focuses on emergent (not selected for) robustness to disruption of evolved multi-cellular entities. Robustness is achieved exploiting dynamic stability, i.e. the continuous replacement of the materials organisms are made of while appearing static. For details of the developmental setup and implementations see chapter 6, where also other approaches that focus on multicellular developmental modelling are discussed in detail.

[Bongard(2002)] evolved GRNs to build body plans together with nervous systems for artificial agents, using a GRN model based on [Reil(1999)] (described above). The creatures first undergo a GRN-controlled growth phase that not only constructs body shape (cylindrical elements linked by joints) but also a neural system. In a second phase, this system controls the organism's movement, which is used to determine fitness. So after the developmental phase the GRN is discarded, as in other approaches for the growth of neural controllers.

[Quick et al(2003)Quick, Nehaniv, Dautenhahn, and Roberts] on the other hand focused on the notion of GRNs as embodied active control systems, with continual coupling of internal and external dynamics. In their *BioSys* model eight GPs (encoded as three bit binary value, '000' to '111') are used. Their separately stored amounts function as state memory (with a certain decay every time step) and are used for input and output: input from environmental sensors increases one particular GP level, output to actuators occurs by reading and interpreting another protein level. GP levels are regulated by a fixed number of genes, with each gene encoded as eight bits – four bits for the binding site (three bit protein that binds to it and one bit to specify whether regulation is activatory or inhibitory) and four bits for the protein produced, three to specify the type again and one for the activation function. Both activation functions have a sigmoid shape, but their offset is shifted: The 'no activation required' function results in GP being produced unless the gene has an inhibitory binding site which actively prevents this, while 'activation required' type genes do not produce GP without activity on an activatory binding site. In the reported experiments a GRN-controlled robot on two wheels is evolved to show light-seeking behaviour ("phototaxis"). Cell-global GP levels are changed / read for environmental coupling. Light to the left of the robot results in input of type zero GP, and correspondingly GP one for light to the right. Similarly, the level of GPs six and seven controls the left and right motor of the robot.

[Taylor(2004)] evolves GRN-controlled robots for a more complex multi-agent clustering task. In the experiment five robots, each with the same genome, have to align in space to achieve a high fitness. The robots have eight 'diffusion sites' (one every 45 degrees), in each of which GP values are stored separately. Genomes are of variable length (initially 1000 characters from $\{0,1,2,3\}$), achieved by a two-point crossover with points chosen independently in the two parents. During parsing, gene sequences are initiated by a gene promoter sequence '010' (so the number of genes is variable, the promoter appears roughly with probability $4^{-3} \approx 0.0156 = 1.56\%$). In the model there are 64 GPs (encoded with three base-4 digits). Accordingly the three characters after the promoter define the type of GP produced by the gene, followed by three digits for the activation function. Similar to the *BioSys* model, the function comes with two offsets (either always on unless inhibited or off until activated), but in addition the shape of the curve can be either sigmoidal or Boolean step-like. The following digit describes where produced GP is placed; either GP is

distributed equally among all eight diffusion sites, or at the site with highest/lowest
GP value or it is placed at a particular diffusion site - in the last case two more digits
specify which. Everything between the end of the gene itself and the next gene promoter is considered potentially regulatory region for the gene starting after it. This
region is scanned for activatory ('12') and inhibitory ('23') promoters. Both signify
the start of a binding site, where the three digits following a promoter determine the
associated GP. To determine GP production the sum of all GPs the gene has binding
sites for is used as input to its activation function. The sum is calculated by adding
GP amounts for activatory and subtracting for inhibitory sites. In this process the
global amount over all eight different diffusion sites is used [11]. The positioning of
these sites features in external interactions: A subset of GPs 'diffuses', which can
be sensed by other individuals as input. Eight GPs (numbers 48 to 55) are produced
at their respective diffusion site if the robot is docking to another robot in that direction. Eight GPs (numbers 56 to 63) are used for sensing depth (produced evenly
among sites). For actuation, the production of 8 GPs (numbers 40 to 47) by the GRN
signal movement in a site's direction.

[Bentley(2004)] also evolves (single) robots for simple tasks like wall-following
and obstacle avoidance. While the mapping from GRN dynamics to robot control is
not very different from the ones described above, the GRN model is novel: a "fractal chemistry". In his approach GPs are square sections (finite square subsets) of the
Mandelbrot set, defined by three real numbers x, y, z – x and y give the centre of the
section and z the the length of the sides. A gene is encoded as several real values,
apart from affinity and amount threshold as well as an output type value, xp, yp, zp,
for the fractal GP binding site, and x, y, z for the fractal GP product. Needless to
say that exact template matching for interaction between GPs and binding sites
would not work with real numbers. Here fractal chemistry helps; interacting GPs
are merged by iterating through their fractal equations in parallel for sample points.
If the length of any of these iterations is unbounded, the associated point will not be
included in the merged GP. This is roughly equivalent to a logical AND. The merged
GP is then compared against the fractal binding site xp, yp, zp to determine whether
the corresponding gene should be activated. For a positive affinity threshold the difference between the fractal shapes has to be equal to or lower than the threshold
for the gene to be activated. For a negative affinity threshold, the difference must be
larger than the absolute value of the threshold for the gene to be activated. While
this matching process is still far away from the intricacies of real GP 3D folding and
interaction, the model takes an interesting approach at capturing more of this.

In summary, it can be said that many parallel developments have led to a large
amount of GRN models, many built for one specific purpose, and their number is
growing more and more. This is why this review can not be exhaustive, but the major developments should have been covered. (Note that multicellular, spatial GRN
models are reviewed in detail in section 6.2.)

Generally, GRNs are modelled statically, with one gene product regulating the
expression of a fixed set of genes with a fixed strength. It seems however that

---

[11] GPs at all diffusion sites decay at the same rate subsequently.

regulatory influences are more dynamic in nature. For example low specificity with varying affinities even make the term regulatory "network" somewhat question-able. Moreover, higher level regulatory mechanisms, such as alternative splicing and post-translational modification are not modelled. Most models know only one level of regulation, with additive regulation, discounting any non-linearities in gene regulation.

3.3 xBioSys Model

A new class of artificial GRNs is introduced. Genes in the xBioSys model are reg-ulated by an evolvable number of complex cis-regulatory control modules, each involving a finite number of inhibitory and activatory binding sites. It allows for complex regulatory logic, with environmental interaction being explicitly consid-ered. This GRN model was first described in [Knabe et al(2006)Knabe, Nehaniv, Schilstra, and Quick] and is inspired by BioSys [Quick et al(2003)Quick, Nehaniv, Dautenhahn, and Roberts]. As in BioSys, cells consist of GPs and a genome. The genome has a fixed number of genes. Gene activation is controlled by regulatory regions organised into *cis-modules* and these in turn contain one or more *binding sites*. Free GPs can "attach" to binding sites. The attachment of GPs to binding sites is restricted by the match of site and GP type[12]. Depending on the attachment of GP to the binding sites the corresponding cis-modules positively or negatively in-fluence the production of GPs. In this model all GPs are potentially regulatory. In contrast to BioSys, in xBioSys there can be any number of cis-modules per gene and every cis-module can have any number of protein binding sites, only limited by the number of unequal crossover events. This is to allow for a second level of pro-tein regulation (absent from previous models), 1) interaction of binding sites within a cis-module and 2) among cis-modules (for details see the description in section 3.3.1.1 below). In logical terms one can think of this grouping of regulatory inputs as an OR of ANDs. The AND level constitutes a canalising function in the sense of [Kauffman(1993)] as one zero value there causes the whole term to be zero no matter what the other stimuli are. The additional control logic level might facilitate the evolution of "master control genes" – active genes at the top of a hierarchy that might start a cascade, turning a large number of other genes, similar to nature's Hox and Pax genes described above. In summary the xBioSys model is trying to come a little closer to nature, where "5-10 regulatory sites are the rule that might even be occupied by complexes of proteins" [Banzhaf(2003)], in an attempt to facilitate the evolution of complex dynamics.

---

[12] Two variants of GP-site matching are used: In template matching a perfect sequence match of binding site and GP is necessary. Smooth matching allows looser matchings like some models described above [Banzhaf(2003), Bentley(2004)]. Details of the matching mecha-nisms are given below.

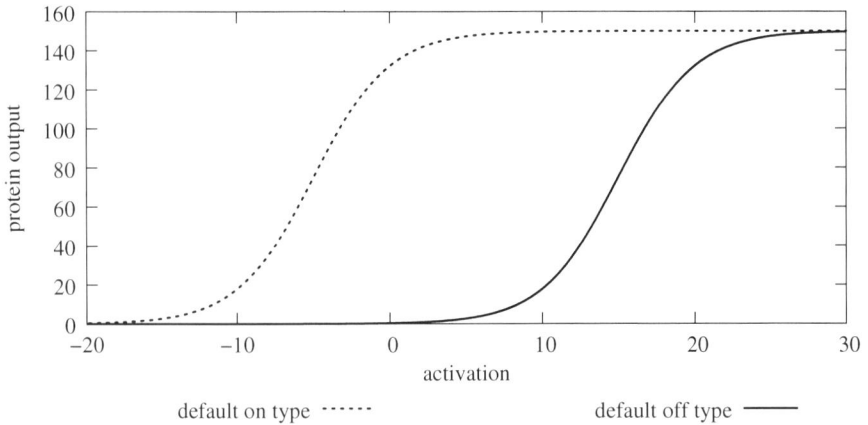

Fig. 3.6 Sigmoid Activation Functions. Every gene produces its proteins according to the cumulative activation level of its cis-modules and its activation type: either even when no activation is present ("default on" - left) or only with positive activation ("default off" - right).

3.3.0.2 Genetic Representation and Genotype-Phenotype Relation

Every GRN's genotype is a string of base four digits, encoding several genes and some global parameters of the corresponding network. Digits 0 and 1 are *coding* digits that may be involved in regulation or protein coding. To differentiate between a sequence of coding bits, a cis-module boundary and a gene boundary the genetic alphabet was increased to four values, with digit 2 delimiting the end of a cis-module and digit 3 delimiting the end of a gene. In the version of the model used here there is a predefined number 2^n of different protein types, so that for example to have eight (2^3) types three bits encode a protein.

In the experiments described here, a fixed number of genes between 3 and 20 was used. After parsing the genome into genes, the last $n + 1$ coding digits of every gene determine its output behaviour, n bits for the protein type produced (by definition a gene has exactly one GP here) and the last bit for the gene's activation type, which can be either *constituitive* ("default on" – active unless repressed) or *induced* ("default off" – silent until activated by regulatory sites), see fig. 3.6. For cis-modules the first coding bit determines its influence on the gene's activation level (*inhibitory/activatory*) and every following n coding digits are considered a protein binding site.

Note that, due to evolutionary operators ("unequal crossover") explained below, there might be additional digits that are not meaningful, e.g. when sites in a cis-module do not equally divide into n-bit regions (where n is the number of bits required to code for a protein). Such digits which are neither translated nor regulatory are referred to as *junk*. See fig. 3.7 for an example gene representation.

Module delimiter

Module   Junk | Expression type

0101110211010200011113

Binding sites   Protein code

Regulator type   Gene delimiter

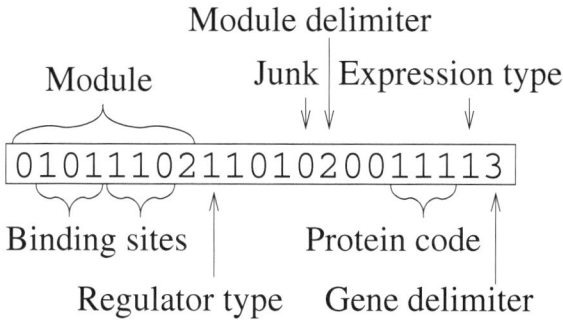

Fig. 3.7 Example gene representation. The gene 0101110211010200011113 (assuming $n =$ 3 bits, $2^3 = 8$ protein types possible) will produce protein 7 (111) and is "off by default" (last digit before the terminating 3 is 1). It has two cis-modules, the first inhibitory (starting with 0) binding a combination of proteins 5 (101) and 6 (110), and an activatory cis-module (starting with 1) to which protein 5 (101) will bind. The last zero of the cis-module 110102 as well as the following two zeros are all ignored, they are "junk".

The genome also encodes several evolvable variables global to the cell. These are 1) the *protein-specific breakdown or decay rates*, four bits for each of the eight protein types, indexing into a fixed look-up table of values, 2) the global *binding proportion*, also four bits indexing into a look-up table, but identical for all proteins, and finally 3) the global *saturation value*, three bits indexing into a look up table the same for all proteins[13]. Look up table values are listed in table 3.1.

Table 3.1 Look-up tables for evolvable variables global to the cell; 1) protein-specific decay rates, 2) binding proportions – "..." means that higher values also map to 1.0 – and 3) saturation values for amount of protein.

1) 0.0	0.1	0.2	0.3	0.4	0.5	0.6	0.7	0.8	0.9	1.0	1.0	...
2) 0.0	0.1	0.2	0.3	0.4	0.5	0.6	0.7	0.8	0.9	1.0	1.0	...
3) 10	25	50	100	150	200	300	500					

3.3.1 Regulatory Logic

The model is run over a series of discrete time steps, its lifetime. In each time step initially a fraction of the free proteins, determined by the global binding proportion parameter, are bound to matching sites. The next two sub-sections describe the two different matching mechanisms compared.

See appendix A for pseudo code of the xBioSys update algorithm to clarify details.

---

[13] The saturation value is the upper bound for the free amount of each protein. To put the values into perspective it should be noted that the maximum amount of protein one gene can produce in one step is limited to $r = 150$, a scaling factor described below.

3.3.1.1 Template or Perfect Matching

Here the fraction of proteins available for binding is assigned to the binding site that has the same binary code as the protein. If there is more than one binding site competing for the same protein the fraction is equally distributed between all matching sites. In this process all protein binding sites are treated equally, regardless of the cis-module to which they belong. Let b_i be the number of all binding sites matching protein i (there can be several for the same protein within one cis-module and, of course, on different cis-modules) and c_i^t denote the amount of free protein i being available for binding at time t. Then the amount p_{ijm}^t of protein i bound at time t to a given binding site in cis-module j of gene m and matching protein i is:

$$p_{ijm}^t = \frac{c_i^t}{b_i} + p_{ijm}^{t-1}, \tag{3.6}$$

where p_{ijm}^{t-1} is the amount of protein i at the binding site in the previous time-step after saturation (upper bound for protein amount) and protein-specific decay (the fraction of the protein amount which is lost in every time step) have been taken into account[14], with the initial condition $p_{ijm}^0 = 0$.

3.3.1.2 Smooth Matching and Specificity Factors (SFs)

Above it was assumed that the binding sites on the cis-modules need to have exactly the same structure (bit pattern) as the protein that binds to them. In this variant proteins might also attach to areas they do not perfectly fit. As a protein's affinity for a binding site should depend on how well they match, a measure of closeness is needed. This is achieved by calculating the Hamming distance between their bit representations. The Hamming distance is simply the number of bits that are different between two strings, so for eight different protein types ($n = 3, 2^n = 8$) the maximum Hamming distance is 3. The distance values are then used as input to a bell shaped function, see fig. 3.8. The width of the curve, i.e. the protein's affinity for binding sites that it does not match perfectly, depends on SFs. SFs were incorporated as additional proteins, one for every of the original 2^n proteins, being produced by genes just as the old ones. So the number of protein types in the cell doubles to 2^{n+1}; however, SF proteins do not bind to binding sites themselves but are only used as modifiers when the other proteins are binding. Every SF influences the binding behaviour of its corresponding protein, modelling protein interactions that modify 3D structure thereby changing binding affinity as described above. Correspondence is defined as SFnumber minus 2^n, i.e. the highest bit of a protein indicates normal protein or SF protein. The bell shaped curve (fig. 3.8) is given by:

$$b(x, \sigma) = \frac{1}{\sigma\sqrt{2\pi}} e^{-\frac{x^2}{2\sigma^2}} \tag{3.7}$$

---

[14] Note that decay rates are fixed over a GRN's lifetime, while in natural GRNs breakdown is also subject to regulation.

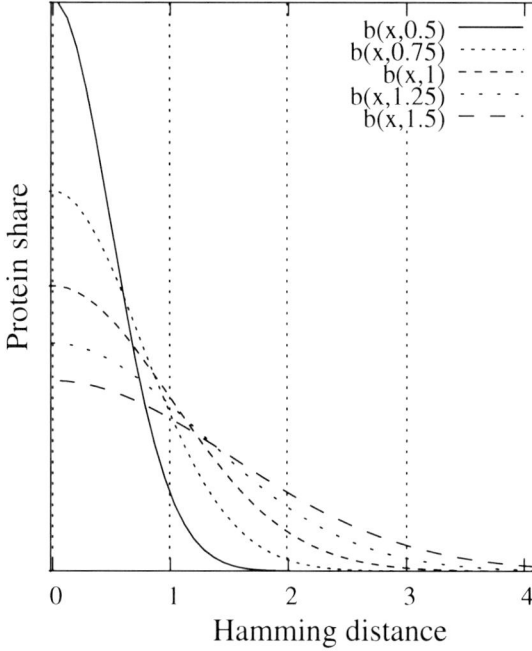

Fig. 3.8 Fraction of proteins that bind to a binding site based on Hamming distance between protein and binding site under influence of specificity factors. The width of the curve is regulated by specificity factors and the dotted vertical lines indicate the possible Hamming distances of the protein's bit representation to a perfect match. There are only four curves shown for clarity, but all intermediates are possible as well. See section 3.3.1.2 for details.

The value for x is the distance of protein i (the one to be distributed) to a binding site and the value for σ is $0.5 + c_i^S$, where $c_i^{S,t}$ is the amount of SF for protein i, $i \in \{0..2^n\}, c_i^{S,t} \in [0,1]$ (SFs are not multiplied by the scaling factor $r = 150$). Let H_{ih} denote the Hamming distance between bit representations i and h. The number of protein types with a distance d can be formalised as $N(d) = |\{x|H_{0x} = d\}|, x \in \{0..2^n\}$.

So for the example with $2^n = 8$ protein types we have $d \in \{0,1,2,3\}$, and for every protein there is one binding site type matching perfectly ($N(d=0) = 1$), three binding site types with $N(1) = 3$, three with $N(2) = 3$, and only one where all bits are different ($N(3) = 1$).

Combined, with the symbols as above:

$$p_{ijm}^t = p_{ijm}^{t-1} + \sum_{h=1}^{2^n} c_h^t \left(\frac{b(H_{ih}, 0.5 + c_h^{S,t})}{\sum_{z=0}^n b(z, 0.5 + c_h^{S,t})} \right) / b_h N(H_{ih}) \qquad (3.8)$$

Note that a) the division by $\sum_{z=0}^{n} b(z, 0.5 + c_h^{S,t}))$ only occurs to normalise the sum of shares to 1, b) proteins decay with the decay–rate specific to the protein (that perfectly matches the binding site they attach to, so one might speak of a binding-site-type specific decay rate now), and c) unlike amounts of normal proteins c_i^t, SFs $c_i^{S,t}$ currently decay completely every time step (i.e. after the above calculations are done). Because of c) no decay rates for SFs need to be evolved and no saturation value applies, but $c_i^{S,t} \in [0,1]$. When calculating the protein output of genes that produce SFs it is simply not multiplied by the scaling factor $r = 150$ (limiting its output to values in $[0,1]$) and if two genes are producing the same SF and their combined output is above 1 the amount value is just set back to 1.

3.3.1.3 Activation Levels

For both matching mechanisms the activation level a_m of gene m with k cis-modules is calculated as

$$a_m = \sum_{j=1}^{k} \pm_j \min_{i:\, \text{protein } i \text{ binds to cis-module } j} p_{ijm}^t, \tag{3.9}$$

where

$$\pm_j = \begin{cases} +1 & \text{if cis-module } j \text{ is activatory} \\ -1 & \text{if cis-module } j \text{ is inhibitory.} \end{cases} \tag{3.10}$$

Table 3.2 AND and min functions are identical for Boolean $\{0,1\}$ inputs

i_1 i_2	$\min(i_1,i_2)$	$\text{AND}(i_1,i_2)$
0 0	0	0
0 1	0	0
1 0	0	0
1 1	1	1

So the calculation of every gene's activation level is done by adding (activatory) or subtracting (inhibitory) the amounts per cis-module but only the lowest value of bound protein per cis-module is used (min). Note that this use of min is an extension of logical AND (see table 3.2) and results in non-additive effects ("synergy") in gene regulation. Furthermore this is a canalising function in the sense of Kauffman [Kauffman(1993)], who underlines their importance for dynamical properties of Boolean networks. For a function to be canalising (at least) one input variable must be able to assume a value that forces a certain output value, regardless of the other inputs – which is clearly the case here as one low input to the min function suffices to ensure a low output. Summing up between cis-modules on the other hand is more akin to a logical OR.

The increase in protein amount due to gene m is then $f_m(a_m)$,[15] where

$$f_m(x) = \begin{cases} \frac{r}{2}\left(\tanh(\frac{x-15}{s}) + 1\right) & \text{if gene } m \text{ is "default off"} \\ \frac{r}{2}\left(\tanh(\frac{x+5}{s}) + 1\right) & \text{if gene } m \text{ is "default on".} \end{cases} \quad (3.11)$$

The parameter $s = 5$ determines the steepness of the slope, with the function becoming more switch-like as s gets smaller, and $r = 150$ determines the range of the function. The output of the gene's activation function is added to the amount of unbound protein of that gene's output protein type. After this calculation the amounts of all unbound proteins are, if necessary, reduced to the global saturation value and then all proteins, free or bound, decay by the protein-specific rate. Finally, environmental input occurs by increasing the unbound amount of certain proteins by some value and output by reading some specified protein amount values. Simple scaling by r is used to map stimulus input levels from the signal range to a protein amount, and *vice versa* for output protein levels. Only GPs used in such a way are structural or functional, but the model does not prevent these from having regulatory effects. See appendix A for pseudo code of the xBioSys update algorithm to clarify details.

3.3.2 Evolutionary Algorithm

A standard Genetic Algorithm with elitism, tournament selection and replacement is used with the GRN representation described in section 3.3 above in all experiments described in this work. Every evolutionary condition was studied with ten runs; each lasting 250 to 500 generations with a population size of 250 individual GRNs. The initial population started with one cis-module per gene and one protein binding site per cis-module, all coding bit values being randomly assigned; in network terms the nodes are randomly connected, with at most one incoming arc.

3.3.2.1 Selection

Later generations are formed by carrying over the best-performing individual of the last generation automatically and, keeping population size constant, the other individuals are replaced by offspring. To generate each pair of offspring, 15 (not necessarily different) individuals of the prior generation are chosen randomly and of these the best two selected to be "parents".

---

[15] For example, for the gene 0101110211010200111113 from above (c.f. fig. 3.7) this would mean that due to the first (inhibitory) cis-module, assuming a share of 20 type 5 proteins (101) and 1 type 6 protein (110) per binding site, the value -1 would go into the sum. The second (activatory) cis-module however would contribute $+20$ resulting in an overall activation of 19, which gives a protein output of about 125 type 7 proteins.

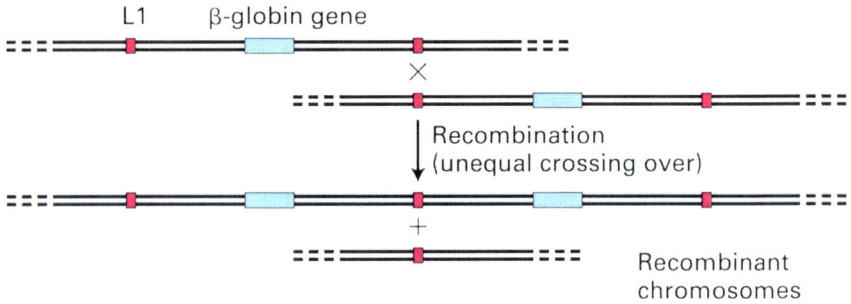

Fig. 3.9 Unequal crossing over in nature – β-globin example. Compare to artificial version, schematically shown in fig. 3.10. From *Molecular Cell Biology*, 5/e by Harvey Lodish, et al. (c) 1986, 1990, 1995, 2000, 2004 by W. H. Freeman and Company. Used with permission.

3.3.2.2 Variability

A (single-point) crossover between the parent genomes occurrs 90 percent of the time and every coding bit is flipped with a mutation probability of one percent. As there can be a variable number of cis- and of protein binding sites per gene genome lengths will vary, so a standard bit-string crossover at the same digit in both genomes could change the number of genes. To conserve all but (at most) one of the genes as basic building units the genomes of the parents are divided into compartments: one compartment for every gene and one compartment for the global variables. Then (with a probability of 0.9) a single compartment is chosen for crossover and in this compartment a point allocated for crossover (no matter how long the two genomes are before that compartment boundary).[16] This process is inspired by the biological mechanism known as synapsis, the pairing of homologous chromosomes where mostly similar sectors pool together. To achieve variable length genes, the unequal crossing-over observed in biology is mimicked: When crossing over from parent 1's genome to the second parent's genome copying does not necessarily continue at the same position of parent 2's genome but is shifted by an offset: Gaussian offset crossover – see fig. 3.10, mimicking the unequal crossing-over observed in biology, see figure 3.9 and [Gregory(2004)].

This offset is randomly drawn from a Gaussian distributed random variable with mean 0 and standard deviation 4. The relatively large number four was chosen to increase the chance of duplicating genetic information, the importance of which was already pointed out by [Ohno(1970)] for the evolution of biological complexity. Ohno put emphasis on whole-genome duplications while it is now, with better techniques, becoming ever clearer that "both small- and large-scale duplication events have played major roles" [Taylor and Raes(2005)].

Note that the offset point is limited to stay within the boundaries of the compartment, hence if crossover point + offset is smaller/larger than the left/right boundary

---

[16] This is why 'at most' one gene is changed: The crossover point could be zero or equal to the gene's coding length.

Gene 1   **Gene 2**   **Gene 3**

1) 1020100113   010111 021 10102001101013   011200113

2) 01102101201113   110111021 101 020011113   1011210200013

3) 1020100113   010111021 ‖ 020011113   1011210200013

4) 01102101201113   110111021 ‖ 02110102001101013   011200113

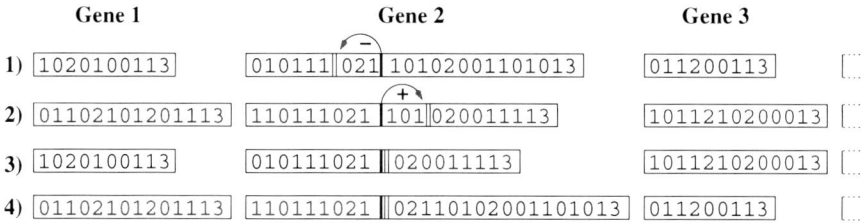

Fig. 3.10 Gaussian offset crossover. Genomes of (1) parent 1, (2) parent 2, (3) offspring 1, (4) offspring 2. Only the compartment chosen for crossover and two neighbouring genes are shown. Both children get digits up to the crossover point (solid bar) from their respective parent, but then continue in the other parent's genome with opposite Gaussian-distributed offsets (−3 and +3, respectively, here).

it is set to the corresponding boundary value. So the number of 2s (cis-modules) might increase by crossover – mutation was only applied to coding digits (0s and 1s) – but not the number of 3s as these are the compartment boundaries. When crossover occurs in the part encoding for global parameters the offset is always set to 0 as offsets would be meaningless here.

These processes allow both neutral crossover and mutational changes, as degenerate cis-modules (i.e. less than three bit – one protein – long) are ignored. Additionally this means that genes could become dysfunctional, in a similar manner to the so-called pseudo-genes found in nature, e.g. if there was not a single cis-module and the gene had an activation type of "off by default".

3.3.2.3 Variations

To check how useful exchange of genetic material in the creation of new individuals is for the algorithm, one experiment was run where only the single best individual of the group of 15 was chosen as parent. Its genome would then be crossed over with itself as described above – so none of the properties of offspring generation mentioned is lost. However chances are then much higher that offspring regulatory dynamics (and thereby performance) are not too different from the parent's as most genes could be crossed only with identical copies of themselves. So in effect the variability has been reduced, likely yielding smoother transitions for this *self-crossover* experiment.

In some experiments gene duplication occurred with a probability of 1 percent per gene, where one gene was chosen randomly and duplicated. When crossover occurred between GRNs with different gene numbers, crossover was bound to happen in compartments up to the number of genes present in the GRN with fewer genes. As early experiments[17] with this gene duplication operator in xBioSys have produced mediocre outcomes it is not used here.

---

[17] These experiments have not been published, but results are available online - at http://panmental.de/CECdynAff/#moreExperiments

Chapter 4
Biological Clocks and Differentiation

[Bonner(2001)] has pointed out the key importance of inventing a stimulus-response system for the evolution of live on earth, in addition to metabolism and replication. As *incessant responsiveness* [West-Eberhard(2003)] and the basis of signalling it has come to be a characteristic of life on earth. Not all organismal responses to external stimuli are simple reactive responses, but behaviour generally depends also on the organism's internal state – and this state can reflect environmental processes as is the case in biological clocks (section 4.2.1) – such internalisation can be evolutionarily advantageous in noisy environments.

In multicellular organisms this stimulus-response system is used to trigger regulation switching between different control dynamics, so-called cell differentiation. As each cell within an individual organism contains an identical copy of the organism's entire genome, it is hard to see how useful division of labour between the cells of an organism could be organised without this mechanism.

From a dynamical systems point of view, bi-stability as in differentiation and oscillations as in biological clocks are basic building blocks of more complex systems [Tyson et al(2008)Tyson, Albert, Goldbeter, Ruoff, and Sible, as well as other articles in special issue].

4.1 Dynamical Systems

This section presents a very short introduction into dynamical systems theory. It can be skipped by readers familiar with basic concepts of dynamical systems. Much more comprehensive descriptions including mathematical definitions can be found, for example, in [Katok and Hasselblatt(1995)] and [Strogatz(1994)].

Any system which has its behaviour determined by a deterministic update rule is a dynamical system. The state of a dynamical system is described by a set of numbers. Given the current state, the update rule determines the following state. The update rule can be an explicit table (if the number of states is finite) or, more often, a formula. Figure 4.1 shows both cases for a Boolean Network with three nodes.

J.F. Knabe: Computational GRNs: Evolvable, Self-organizing Systems, SCI 428, pp. 45–70.
springerlink.com

1 — 3 AND 2

1 NXOR 3 — 2

3 — 3 IMPL 2

t=n 1 2 3	t=n+1 1 2 3
0 0 0	0 1 1
0 0 1	1 0 0
0 1 0	1 1 1
1 0 0	0 0 1
0 1 1	1 0 1
1 1 0	1 0 1
1 0 1	1 1 0
1 1 1	1 1 1

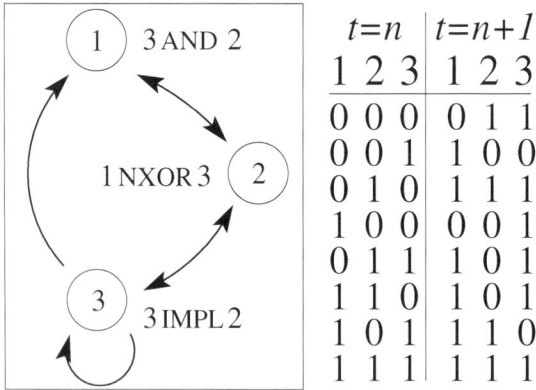

Fig. 4.1 Example state update table (right) of a Boolean Network (left, shown and explained in fig. 3.5). Given a state of the nodes at time $t = n$ the system will move to the state shown at $t = n + 1$.

System states can be represented by a point in a state space with the appropriate number of dimensions. Unlike in the Boolean Network example, for continuous systems described by differential equations, "time steps" and numeric changes are usually infinitesimally small. Given an initial state the update rule can be applied repeatedly, resulting in a trajectory (or orbit) through state space. Small changes can have big impact in the long-run in complex systems with feedback – Minimal differences accumulate over time, making prediction hard to impossible as initial measurements might never be exact enough (the famous "butterfly effect"). This is one reason why long-term weather forecasts are never perfect.

Trajectories may return to a state repeatedly and have to do so at some point for finite systems. Given the Boolean Network and update table shown in figure 4.1, from an initial state of 010 (node 1 is inactive, node 2 is active, node 3 is inactive) the system would change to 111 and stay in that state forever - a fixed point. From a state of 000 however the system would move to 011 and further to 101, from which it would change to 110, afterwards always switching between 101 and 110 - a limit cycle (with a very small period). For the last example, in the state 101, turning the first bit off will result in a state of 001. The system has been perturbed and is now in another attractor, cycling between 001 and 100.

Fixed points may be attractors, i.e. once close enough the system will invariably move towards it, whatever the particular trajectory was. The whole "close" area is called basin of attraction. Of course not only fixed points can be attractors but any set of states, as long as the system will remain on the attractor. Two types of attractors in a three dimensional state space are shown in figure 4.2.

To make the connection to this chapter's main topic: The prevalent opinion amongst researchers is, that periodic oscillations as in biological clocks can be viewed as limit cycles, while bi-stability as in differentiation corresponds to switching between basins of attraction.

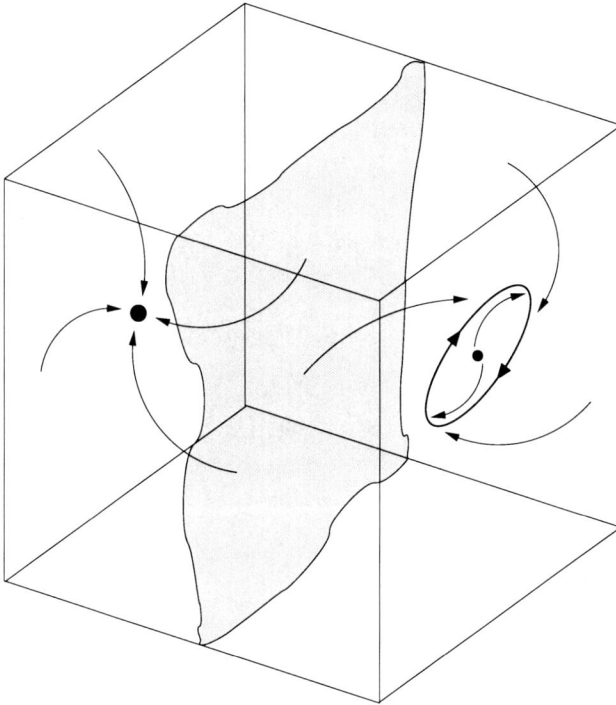

Fig. 4.2 Three dimensional state space with two basins of attraction, separated by a wall. States on either side of the wall will end up the corresponding basin of attraction after following a deterministic path, unless strongly disturbed. On the left a stable fixed point attractor is shown while on the right an oscillatory limit cycle is shown. After [Kauffman(1993)].

4.2 Biological Background

4.2.1 *Biological Clocks*

Biological clocks provide one of the simplest yet most characteristic examples of internalised incessant responsiveness for life as it has evolved on the earth in that an organism's regulatory dynamics respond with periodic activity in close coupling with periodic cycles of environmental stimuli as experienced in the rhythm of light and dark, or in the effects of lunar gravitation in the ebb and flow of tides.

Biological systems exhibit periodic behaviour on different time scales, but circadian rhythms are believed to have originated already in the earliest cells. The atmosphere on early earth allowed much higher ultraviolet radiation to penetrate during day-time, and nightly cell division provided protection for replicating DNA. This replication behaviour can, for example, today still be found in the single-celled organism *Gonyaulax polyedra* [Winfree(1986)].

Without the capacity to adjust to external signals, minute differences in timing period soon accumulate, leading to internal clocks being hopelessly out of step with the environment.

Following [Winfree(1980), Winfree(1986)], one may ask, How is it that biological clocks still work when external stimuli are hidden (like the sun or other temporal cues) in isolation experiments on living organisms? How is it that they can adapt, within limits, to perturbations in cycle length, phase shift, and resetting? Why in isolation do they run at rates somewhat different from that of the external cycles (with rates being species specific; while the aforementioned *Gonyaulax* has an internalised rhythm of roughly 23 hours, humans are closer to 25 hours)? Are these accidents of neutral selective value, or do they have some adaptive significance at the individual (or lineage) level?

In evolutionary and developmental biology, internalisation of environmental stimuli ([Waddington(1953)] genetic assimilation[1] and the more general Baldwin effect[2]) provides robustness and adaptation to environmental perturbations experienced by a population over evolutionary time.

4.2.2 Differentiation

In every multicellular organism on earth, almost all of an individual's cells contain the same genome but still, depending on signals or differences in the internal chemical composition, can take on very different functional roles [Jacob and Monod(1963)]. The most crucial signalling is believed to be induced by other cells or the environment early in development, e.g. turning on homeotic genes, which "remain on through adult life and maintain particular aspects of the pattern of gene expression characteristic of that segment [they are part of]" [Ptashne(1992)].

There are basically two ways differentiation is achieved in natural cells: Autonomous and conditional. Cell intrinsic properties can give rise to autonomous differentiation during cleavage (cell division), for example asymmetrically distributed gene products. Maternally produced molecules play an important role in breaking the symmetry during the first cell divisions. Some of these so-called maternal factors accumulate at the dorsal side of the egg cell, leading to an unequal concentration distribution after cleavage.

Conditional differentiation relies on extrinsic cues from concentration gradients of molecules (morphogens) or other cells. A third type, called syncytial

---

[1] Genetic assimilation is a process by which species that get a selective advantage from responding to some kind of environmental stimuli evolve to produce the response even in absence of the stimulus. In a classic experiment, Waddington exposed fruit fly embryos to ether, causing deformation to various degrees as canalisation broke down. After artificially selecting for strong deformations of a particular kind for 20 generations, the deformation was even present when embryos were not exposed to ether. Also see section 2.2.

[2] The Baldwin effect refers to scenarios where a behaviour or trait that results from the interaction of organisms with the environment is gradually incorporated into the the species' genetic repertoire [Baldwin(1896), West-Eberhard(2003)].

differentiation, is more a special case of conditional differentiation: In some species, for example *Drosophila melanogaster*, the egg cell undergoes incomplete divisions up to a stage. The nuclei replicate but not the plasma membrane, leading to many GRNs in one cell membrane (syncytium). GRNs in the nuclei can react to morphogen differences and differentiate, but they do so within a single cell membrane.

4.3 Evolving Biological Clocks

The evolvability and dynamics of artificial GRNs, as active control systems, realising simple models of biological clocks that have evolved to respond to periodic environmental stimuli of various kinds with appropriate periodic behaviours, is studied. This was inspired by the fact that internalised biological clocks are found already in the most simple single-celled organisms [Winfree(1986)] and should not be thought of as a cell's full regulatory repertoire but as a module in a larger active network, e.g. as internal timer; as "Transcription can change within one cell type in response to an external signal or in accordance with a biological clock; some genes, for instance, undergo a daily cycle between low and high transcription rates." [Lodish et al(2004)Lodish, Baltimore, Berk, Zipursky, Matsudaira, and Darnell, p. 23].

Previous work on biological clocks in nature has noted the capacity of clocks to oscillate in the absence of environmental stimuli, putting forth several candidate explanations for their observed behaviour, related to anticipation of environmental conditions, compartmentation of activities in time, and robustness to perturbations of various kinds, or unselected accidents of neutral selection. Several of these hypotheses are explored by evolving GRNs with and without (Gaussian) noise and "black out periods" for environmental stimulation. Robustness to certain types of perturbation appears to account for some, but not all, dynamical properties of the evolved networks. Unselected abilities, also observed for biological clocks, include the capacity to adapt to change in wavelength of environmental stimulus, and to clock resetting.

In related work, [Drennan and Beer(2006)] evolve a postulated model of a repressilator; a network with three cyclically-inhibiting elements to produce oscillations, however there oscillations had very short periods and the task was not to oscillate at any particular period.

4.3.1 Environmental Coupling

As stated earlier, environmental cycles have a huge impact on the life of organisms on earth. But in what way these stimuli affect an active organism via its signal transduction pathways and what behaviour is appropriate depends on the type of organism. Here evolutionary conditions are systematically varied by changing the periodic pattern of external signal received at the cellular level — in some scenarios distorted or interrupted or both — as well as the periodic output behaviour expected. As a control condition it was also tested what performance could be achieved without any external signals ever. Experimental variations included whether input and target output were in the same or fixed shifted phase, and whether GRNs were started in a random phase (Phase is the fraction of a complete cycle an oscillation is in). Afterwards we should be able to understand better how distorted or sparse stimuli during evolutionary history affect evolvability. Furthermore evolved individuals can be analysed, e.g. for features like internalisation and clock resetting.

Input stimuli: As described above, input generally was realised as changing a pre-defined protein level every time step. Note that input and output proteins where chosen as far apart as possible in their bit representation (e.g. 0 and 7 for the three bit experiments). This does not matter when using the perfect matching variant of xBioSys, but when smooth matching is used interference would be possible for "close" protein types (with similar bit pattern).

For these experiments, the basic idea was to have periodic environmental stimuli based on a sine curve (shifted to the interval $[0, 1]$). The wavelength wl was set to 20 time steps, while the lifetime for every GRN was 400 steps. Formally:

$$stim(p) = (\sin(\frac{2\pi p}{wl}) + 1)/2 \qquad (4.1)$$

Variations included having only the positive part of sine, a periodic step function, and a brief pulse. The four functions used are depicted in fig. 4.3. In addition, it was varied whether Gaussian noise or "black-outs", periods of no external signal, were applied, yielding four further conditions: $[\pm noise, \pm blackout]$. In $[-noise, -blackout]$ scenarios, the input signal was transduced to yield a corresponding input of a particular protein as described above, without any distortion. In a $[+noise]$ condition, Gaussian white noise with a standard deviation of 0.1 was added to model imperfect signal transduction.[3] For $[+blackout]$ conditions, at random points in time the input stopped completely for an interval of time: every GRN experienced two periods at random times without input, each lasting for 5 percent of its lifetime. In the $[+noise, +blackout]$ scenarios, these perturbations were combined (with the black-out being stronger than the noise, so there was no input – not even noise – during black-out periods).

Output behaviour: Two periodic target functions were used to measure the performance of an individual and assign fitness: sine (fig. 4.3 1) and step (fig. 4.3 3). As stated above, the target output was never distorted by noise and blackouts. Its shape and phase might differ from the input, however the wavelength was identical to the input. Fitness was measured based on the deviation from this target output, i.e. the smaller the value, the better adapted the GRN.

Letting $c_{i_0}^t$ denote the (unbound) amount of the GRN's output protein i_0 and d^t the target output at time t the overall deviation is simply calculated as: $D = \sum_{t=1}^{L} |c_{i_0}^t - d^t|$. The lifetime L of every individual was set to 400 time steps. A randomly-generated initial GRN could typically achieve a deviation of

---

[3] Note however that values below zero are set to zero as negative protein input is not possible.

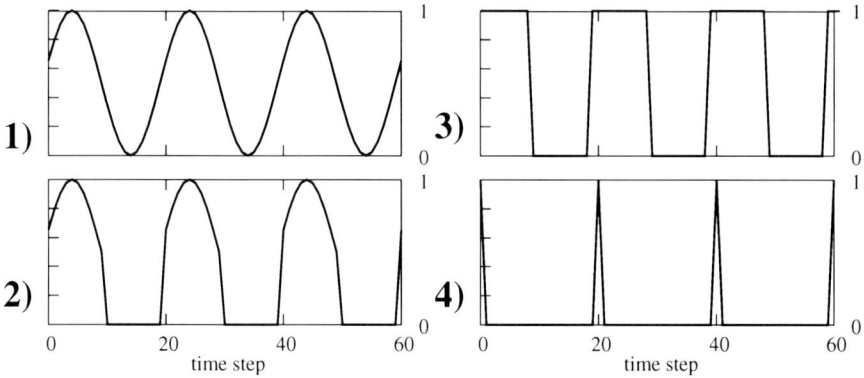

Fig. 4.3 Periodic functions used: 1) sine, 2) positive part of sine, 3) step, 4) pulse

approximately 200. Finally, this value was used to transform the deviation to a standard 0 to 100 performance scale: $(200 - D)/2$, so zero deviation would result in a perfect performance value of 100.

4.3.2 Experimental Scenarios

Overall 32 evolutionary conditions were tested (two target output types times four environmental stimulus input functions in four environmental coupling variations each, as described above) and every experiment was run ten times. Additionally the number of genes was varied (see below). To test the results for robustness, the complete set of experiments was run in four variations denoted as follows:

$[0]$: Always starting in a fixed phase, with no shift between input and target output phase ($d^t = stim(t)$).
$[\frac{1}{2}]$: Starting in a fixed phase, but with a phase shift of $\frac{1}{2}$ period between input and target output ($d^t = stim(t + \frac{wl}{2})$).[4]
$[r]$: Starting in a random phase for each individual, with no shift between input and target output phase ($d^t = stim(t + R), R \in [0, wl]$).
$[\frac{1}{2} + r]$: Combining $[\frac{1}{2}]$ and $[r]$, i.e. shifting the input's phase by half the wavelength plus a random number while shifting the output's phase by that random number only ($d^t = stim(t + R + \frac{wl}{2}), R \in [0, wl]$).

Furthermore, $[0]$ denotes no environmental input ever and $[s]$ the self-crossover condition.

To test how the evolved GRNs were affected by their evolutionary history the best ones were also put into environments not experienced by them or their ancestors before. At first, they got perturbed stimuli, i.e. variations of their usual input

---

[4] Experiments with phase shifts of $\frac{1}{4}$ and $\frac{3}{4}$ have also been run, but as they are not qualitatively different the results are only given online, at http://panmental.de/GRNclocks.

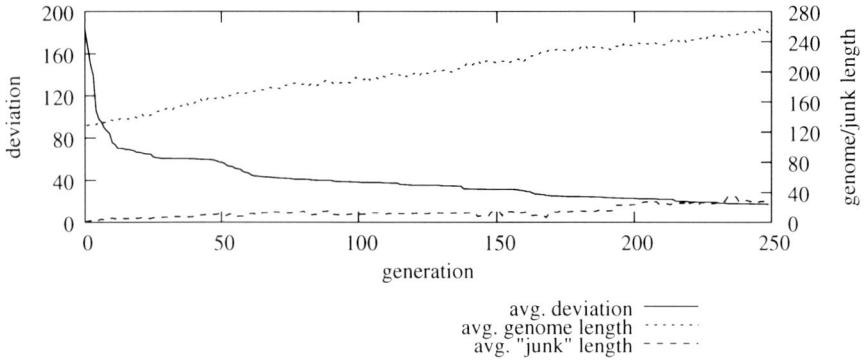

Fig. 4.4 Typical course of an experimental run. Averages are over ten GRNs, each being the generation's best performing GRN in its respective run.

functions with noise and/or black-outs unlike the evolutionary history of their lineage. Afterwards special new stimuli were used: constant input, phase shifted input, different wavelength input, or very long blackout periods.

4.3.3 Results

In almost every single run well adapted GRNs evolved, see table 4.1. Several individuals even achieved deviation below 1 in some of the $[-noise, -blackout]$ conditions.

A selection and summary of the most important results is presented here, detailed tables for all experiments, with lifetime graphs of GRN behaviours can be found at http://panmental.de/GRNclocks. The first subsection examines how populations evolved and the average performances achieved; here a general comparison between different scenarios is made. In the second subsection the focus of investigation shifts to the individual regulatory dynamics, especially as observed in well adapted GRNs.

4.3.3.1 Evolutionary Dynamics

Due to junk as well as inactive genes and binding sites, neutral changes could be observed, i.e. despite the fact that performance often stayed the same for some evolutionary period, genome length might change during crossover, or bits without function might be flipped. Although when crossing over with an offset different from zero usually both a shorter and a longer descendant are produced, the average population genome length increases over evolutionary time. The amount of junk also increases, though at a slower rate, see fig. 4.4 for data.

Table 4.1 Overview of performances, with the topmost row showing the target output behaviour for every run and the leftmost column indicating sample environmental stimuli. Four evolutionary conditions with varying stimuli reliability are compared per data cell. Per condition the final performance of the best individual, averaged over 10 runs with 250 generations each, ± its standard deviation, as well as the best ever observed GRN's performance are shown. Results are for 9 gene GRNs with input and output being in sync, but with a random start phase for every GRN.

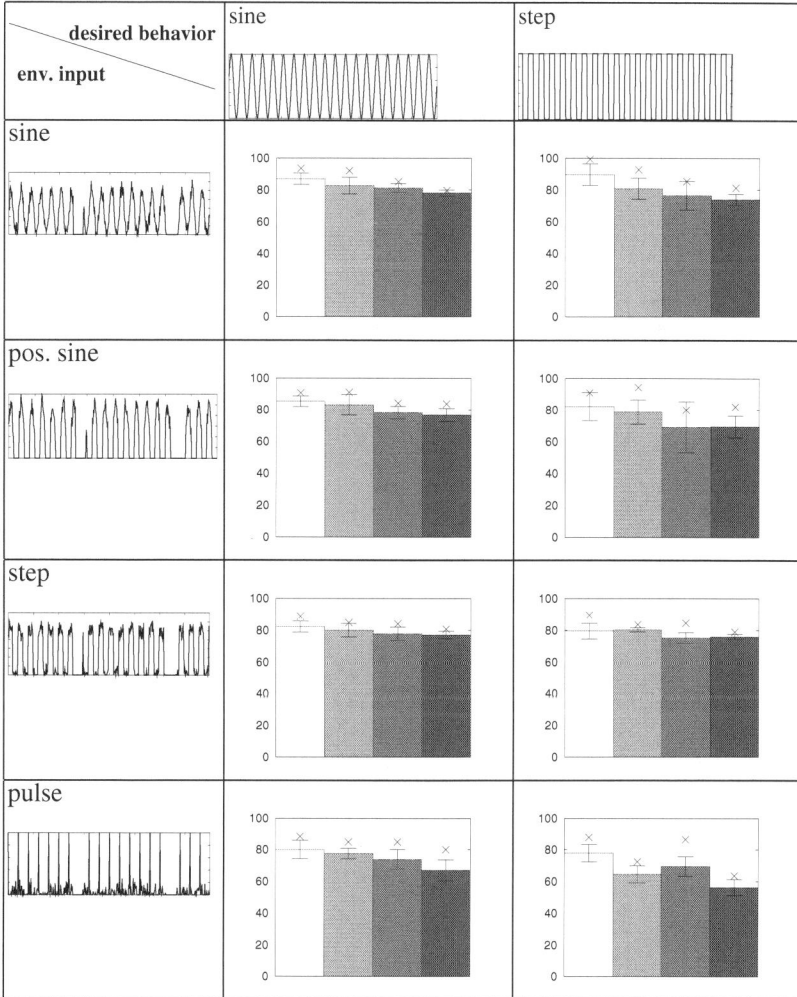

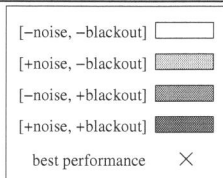

Table 4.2 Outcomes of $[-noise, -blackout]$ runs, with the leftmost column depicting the environmental stimuli used and the topmost row the target output behaviour for every run. Eight evolutionary conditions are compared per data cell, apart from where stated otherwise, results are for 9 gene GRNs always starting in the same phase. Per condition the final performance of the best individual, averaged over 10 runs with 250 generations each, ± its standard deviation, as well as the best ever observed GRN's performance are shown.

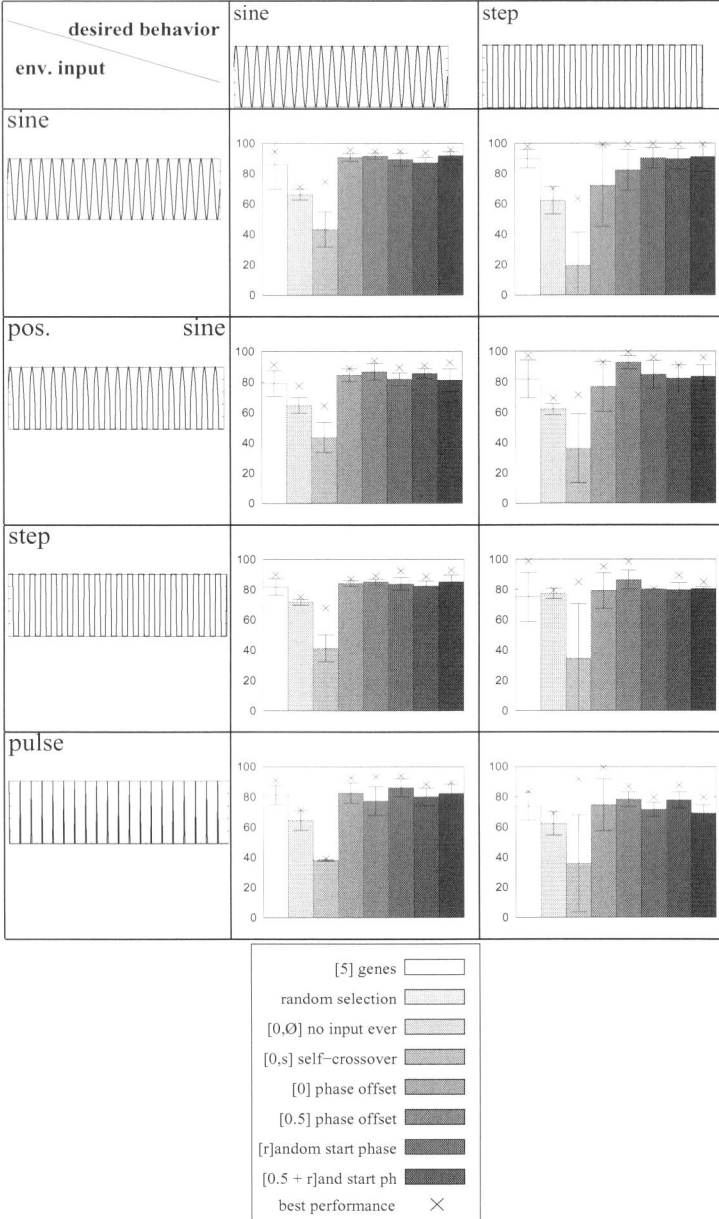

desired behavior / env. input	sine	step
sine		
pos. sine		
step		
pulse		

Legend:
- [5] genes
- random selection
- [0,∅] no input ever
- [0,s] self−crossover
- [0] phase offset
- [0.5] phase offset
- [r]andom start phase
- [0.5 + r]and start ph
- best performance ×

Running the same set of experiments with a fixed number of 5 genes and with 9 genes it turned out that the ones with 9 genes in most cases ended up with a performance superior to their 5-gene equivalents (cf. table 4.2). In the tables and in the following text always refer to 9 gene GRNs if not mentioned explicitly, 5 gene results are available online on the web page mentioned above.

Examining the $[0, \emptyset]$ results shown in table 4.2 it is not hard to see that GRNs are at a big disadvantage when they never get any input stimuli from the environment – so one might conclude that it is hard or even impossible to find regulatory dynamics that do not rely on external information to a certain degree. However in one experiment a GRN was identified that had access to external stimuli but ignored it while still performing well (see below for details). Knowing this it was found that, when well evolved GRNs from conditions with input were evolved a little further in an environment without input, some of them were able to return to a performance level similar to the one they had when input was present within 10 generations. Apparently some "perception" of the environment by the lineage allowed access to areas of the evolutionary search space hardly accessible otherwise.

Comparing the $[0]$ with the $[0, s]$ condition results, self-crossover performs worse in 7 out of 8 experiments. At least for the problem at hand it can be concluded that higher variability due to the mixing of genomes outperforms presumably smoother self–crossover. Of course this will always depend on the structure of the search space and how exploitative exploration shall proceed.

As mentioned before, in the reference case $[0]$ with no phase shift and 9 genes, final GRNs performed reasonably well, and at least one GRN did not use environmental input at all. With shifted input stimuli, condition $[\frac{1}{2}]$, the outcomes were generally comparable to the reference condition. So inverted patterns can be easily created as output and performance is almost as good as without the phase shift. When starting in a random phase (condition $[r]$) well adapted GRNs were of course forced to use external stimuli at least once to synchronise with the environment – but still the rhythm (usually with a shorter period, though) was internalised in most cases. The resulting performance was only slightly worse and in some experiments even better than in the reference condition $[0]$. The same holds for a combination of the above variants (condition $[\frac{1}{2} + r]$).

4.3.3.2 Evolved Regulatory Dynamics

In all scenarios evolved GRNs exhibited a close match to the target output profile and almost always relied on external signals to produce this behaviour. As an exception, the best performing 9-gene GRN evolved with pulse input (fig. 4.3 4) distorted by noise and black-outs to produce a step output had no binding sites for the input protein, i.e. it did not rely on environmental stimuli at all. Apparently the regulatory logic was in principle able to generate a close match to that target output without any external stimuli when every individuals' lifetime always began at exactly the same phase. However the evolution of such dynamics was rare and it is probably not a coincidence that this happened under the evolutionary conditions where only the starting phase was reliable.

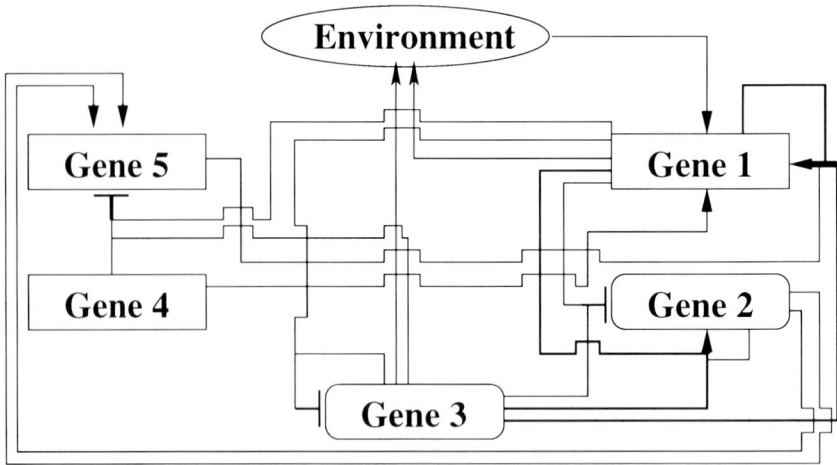

Fig. 4.5 Regulatory interaction diagram of a evolved 5-gene GRN. Boxes denote genes (rounded corners indicating "default on" ones with the others being "default off"), connections ending in an arrow are for activatory influences and the T-like endings depict inhibitory ones. Bold connections mean that the target gene has two binding sites for this protein type and will thus bind a bigger share thereof.

Complex interaction networks evolved and all the best evolved GRNs made use of AND-like regulatory logic with several binding sites bundled to a cis-module as described above (an example of a 5-gene GRN is shown in figure 4.5), although the initial random nets started with only one site per module.

4.3.3.3 Heterochrony

Smooth evolutionary regulation, i.e. changes in the timing of gene expression without affecting the general dynamics, is achieved by varying protein decay rates and the binding proportion (cf. fig. 4.7). The achievability of similar phenomena in GRN models has also been demonstrated in [Banzhaf(2003)]. Its importance for evolution in the timing of developmental control, usually more specifically referred to as heterochronic control, has been appraised by biologists, see e.g. [Gould(1977)], [Buss(1987), p. 105].

4.3.3.4 Phase Resetting

Again inspired by [Winfree(1986)] work the phase resetting behaviour of the evolved GRNs was also analysed. Organisms, deprived of their usual environmental clues to phase (e.g. dark room), usually follow some internalised rhythm instead. This however can be disturbed by giving stimuli (e.g. flash of light) – the rhythm continues afterwards, but may be at another point in its cycle. Systematically done,

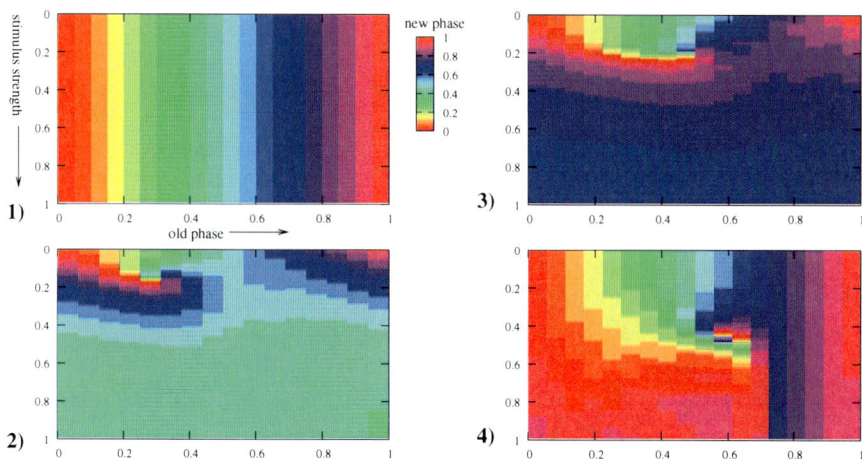

Fig. 4.6 Phase resetting plots. Examples from 9 gene GRNs with internalised rhythms of wavelengths 16–20. 1) Plot of a network that does not respond to input stimuli. The GRN's phase is not changed, i.e. old equals new phase (the new phase is colour coded). 2)–4) Applying stimuli of a certain strength results in a shift in the phase these GRNs are in. Small stimuli have almost no effect, large stimuli reset immediately regardless of the old phase. For particular areas a huge variety of new phases can be achieved with small differences in stimulus strength. All GRNs taken from evolutionary runs without phase shifts or random phases ($[-noise, -blackout]$) but are robust to resetting. See text for details.

one can for an interval of old phases the organism is in, apply stimuli of different strength and record the resulting new phase. To map this data on a 2D plot, the axes are for old phase and stimulus strength, while the important representational trick is to code the new phase in colour. As the colour circle is cyclic (without sharp boundaries or discontinuity) it can be repeated smoothly like a wave. For the GRNs with internalised rhythm (see next subsection), instead of exposing them to some periodic stimulus, an input with a length of five time steps was applied at a particular phase of their cycle (old phase) and then the new phase recorded[5]. While resulting phase resetting plots showed diverse structure, some of the features typically exhibited are discussed. These are illustrated by example plots, see fig. 4.6. Easy to understand is the behaviour with a very weak stimulus: the phase stays the same, no change was observed. For very strong stimuli, the internal oscillator is harshly reset – a shift to one particular phase was found, regardless of the phase the system was in before the input stimulus was given. Most interesting is what was found in between for medium strength stimuli – surprisingly there is not always a smooth gradient between these extremes: It was often observed that in a small stimulus strength interval from a

---

[5] Internalised rhythm was identified by searching for regular peaks in the output protein in the absence of input stimuli. Phase is the fraction of a complete cycle an oscillation is in. The new phase was measured after a settling time of more than 1000 time steps. Source code is available at http://panmental.de/GRNclocks/#code.

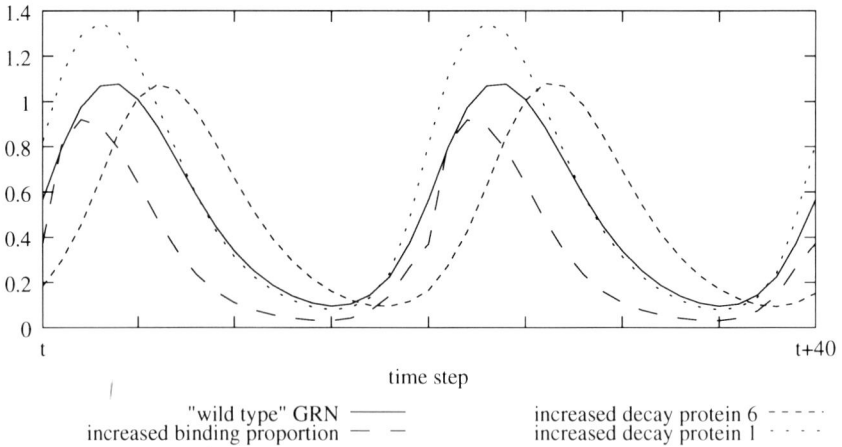

Fig. 4.7 Smooth evolutionary regulation is achieved most easily through variable decay and binding rates. Shown is the behaviour of a 9-gene GRN that is well adapted to producing sine waves (the evolved "wild type") and slight variations of it. The variations' genomes are at most two bit flips away from the wild type.

particular old phase almost every new phase could be reached. Given continuous regulatory dynamics and environmental input, such a singularity necessarily has to occur for mathematical, topological reasons [Winfree(1986), ch. 4].

4.3.3.5   Behaviour in Evolutionarily New Conditions

In what way and how strongly the evolved GRNs relied on input from the environment turned out to depend strongly on the conditions under which they evolved (the evolutionary history of their lineage).

Internalised Rhythms. When input stimuli were hidden, GRNs often exhibited an internalised output wavelength different from the one which was the target during evolution, with some behaviours being quasi-periodic, i.e. almost but not exactly retracing their paths through phase space. Such behaviour occurred mainly in GRNs evolved under pulse input, where the systems used the occasional input to stay synchronised, see fig. 4.8. Apart from fast oscillations systems with internal periods of varying length from 16 to almost 50 time steps were observed (fig. 4.9.1).

Unselected Dynamical Properties. Like most biological clocks studied in man and nature [Winfree(1986)], nearly all the best evolved GRNs – except those that completely ignored their input (which arose seldom and in only one scenario) – in the various scenarios were robust to the shifts in phase and limited changes in wavelength of periodic environmental stimuli. Wavelengths of 19 or 21 mostly did not pose problems, only higher deviations from the normal cycle length of 25% or more

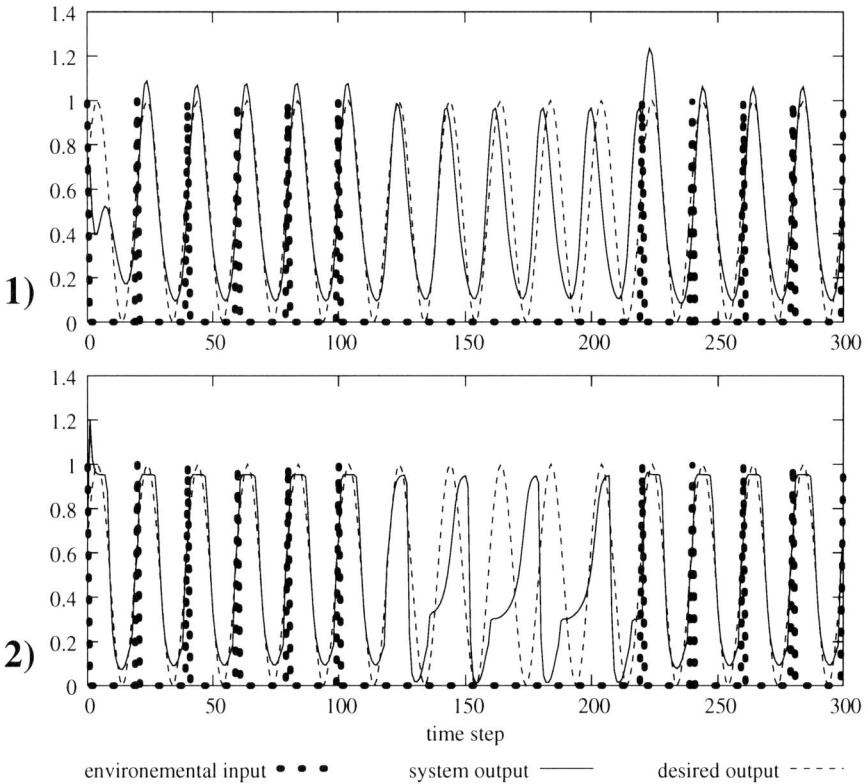

Fig. 4.8 Periodic Behaviour during Blackout. Plot of the output behaviour of GRNs with 1) nine 2) five genes. There is a pulsed input every twenty steps, however from time step 100 to 200 environmental input is suppressed completely. Target output refers to what behaviour was required during evolution, here sine output. Note how the GRN in 1) gets slowly out of synchrony with the desired sine output during the blackout period while this happens quickly for the one in 2). Both GRNs are the result of evolutionary runs with shorter blackout periods of only 20 time steps and achieved a similar performance. The interaction diagram of GRN 2) is shown in fig. 4.5 and variations of the behaviour of GRN 1) are shown in fig. 4.7.

led to distorted outputs. Shifts in phase, i.e. when the input stimulus immediately jumped to another point of the cycle, were tolerated by the evolved GRNs without trouble at all. This occurs despite their lineage never having experienced such perturbations, i.e. without any selection for these capabilities.

When GRNs evolved without noise (and/or black-outs) were placed in an environment with noise (and/or black-outs), performance was still good, however – unsurprisingly – always worse than that of GRNs evolved for such environments.

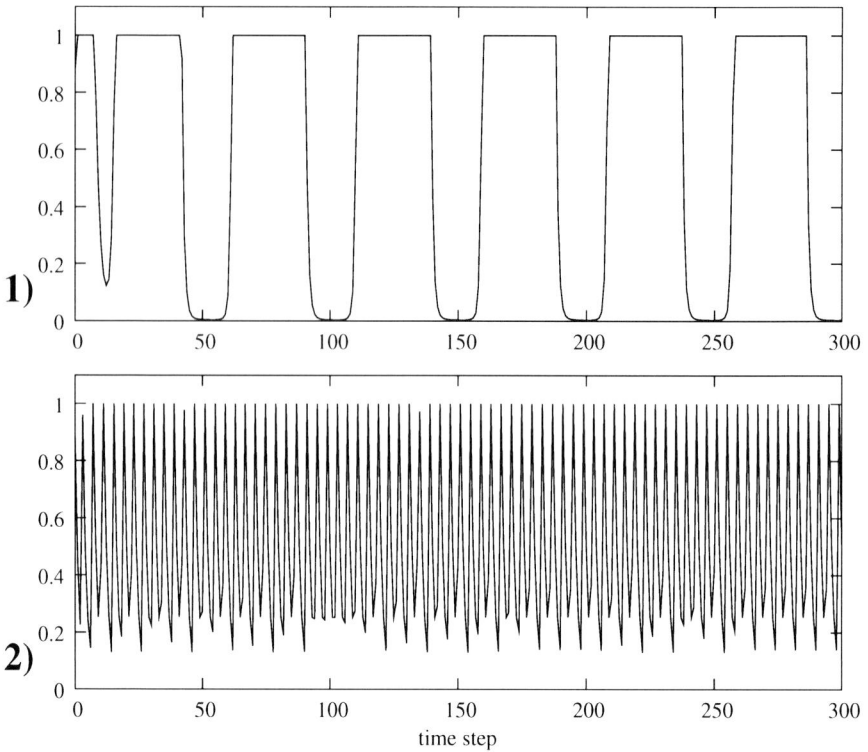

Fig. 4.9 Behaviour without any external stimulus. Plot of the dynamics of GRNs with 1) nine 2) five genes when there is no input ever. The internal period (if it exists) can be very different from the one the environment usually imposes. Both GRNs are the result of evolutionary runs using pulse input, $[-noise, -blackout]$, with step output desired and were the best of their runs.

4.3.4 Discussion

It has been shown empirically that xBioSys GRNs can easily by evolved to exhibit cyclic behaviour, generally in response to periodic stimuli, like that of biological clocks in nature. Moreover, like natural biological clocks the evolved regulatory dynamics of artificial GRN clocks tend to be robust to non-selected perturbations such as phase shift, small period changes, and so on, as well as to perturbations that have occurred in the lineage's evolutionary history (such as noise and blackouts). Especially when there is a sparse signal and lower reliability in environmental stimuli, it pays the GRNs to internalise the rhythm. To what extent the evolved GRN relies on environmental input depends on the character of coupling to the environment during evolution of its lineage.

When evolving without any sensory stimuli from the environment (see results for $[0, \emptyset]$ in table 4.2), it was very hard to find reasonably well performing GRNs. On the other hand, everything else being unchanged, periodic dynamics that did not need input could be found when stimuli were present at an earlier evolutionary stage. This indicates that coupling with the environment made regions of the evolutionary search space accessible that are difficult to reach otherwise.

The importance of crossover in evolutionary algorithms has been debated, as resulting offspring phenotypes might be not very similar to either of their parents. However for this problem the higher variability and thereby sampling of the search space out-competed self-crossover ("both parents identical").

4.4 Evolving Differentiation

Understanding the evolvability of simple differentiating multicellular systems is a fundamental problem in the biology of genetic regulatory networks, and in computational applications inspired by the metaphor of growing and developing networks. Here, the target of the evolutionary algorithm (EA) is a simple two-celled model of differentiation in the sense of Jacob and Monod [Jacob and Monod(1963)], who defined that "two cells are differentiated with respect from one another if, while they harbour the same genome, the pattern of proteins which they synthesise is different". Cell cleavage and development are subject to abstraction; from the start there are two identical cells receiving the same periodic external stimuli, cf. fig. 4.10. The expected difference in behaviour is only signalled by a type inducer (one raised protein level), which can be thought of as being the result of either an internal gene turned on in one cell only during cell division[6] or an externally generated developmental signal.

4.4.1 Environmental Coupling

As in the coupling for the biological clock experiment, c.f. section 4.3.1, evolutionary conditions are systematically varied by changing the pattern of external signal received at the cellular level as well as the periodic output behaviour expected.

Input stimuli: The basic idea again was to have periodic environmental stimuli based on a sine curve (shifted to the interval $[0, 1]$). The wavelength w was set to 20 time steps, while the lifetime L for every GRN was 400 steps. Variations included having only the positive part of sine, a periodic step function, and a brief pulse. The four functions used are depicted in fig. 4.11. The impact of perturbations like Gaussian noise and black-out periods in the input during evolution was investigated in section 4.3.

---

[6] Anterior-posterior cell polarity can result in unequal distribution of proteins after division, which in turn can lead to different gene expression patterns [Wolpert et al(2007)Wolpert, Jessell, Lawrence, Meyerowitz, Robertson, and Smith].

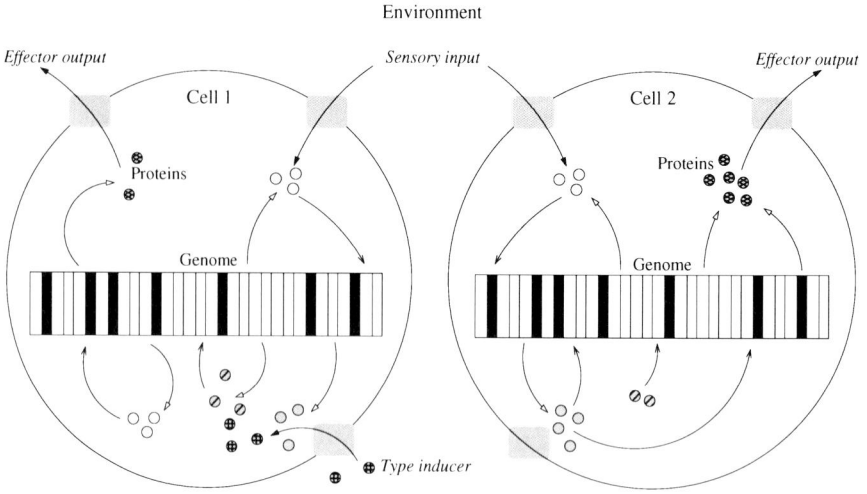

Fig. 4.10 Schematic drawing of the differentiating genetic regulatory network model. The two cells of the simple multicellular individual have the same genome and thus the same regulatory network but can produce very different behaviour, induced by a very simple signal which is here shown as external, but it could also be an internal gene that is always active, e.g. due to cell division disparity resulting in an unequal distribution of proteins or other factors after cell cleavage.

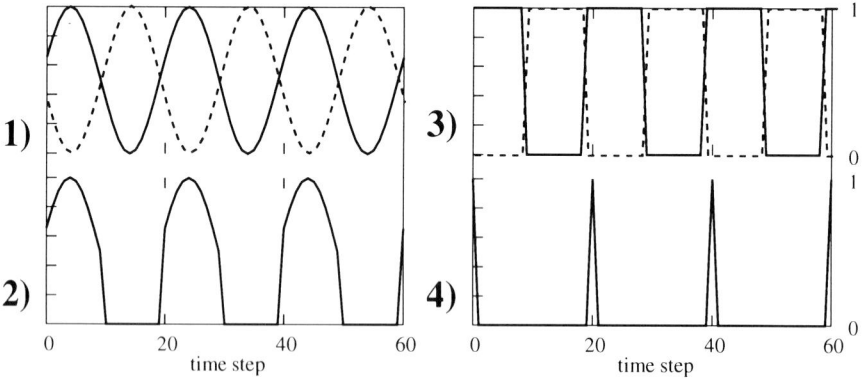

Fig. 4.11 Periodic functions used: 1) sine, 2) positive part of sine, 3) step, 4) pulse – as in figure 4.3, but additionally the inverse or shifted wave was used (for 1 and 3, dashed line).

As mentioned above, both cells of an individual always received the same periodic stimuli, however one cell additionally received an *inducing* signal at saturation value (see scaling, $r = 150$).

4.4.1.1   Output Behaviour

Two periodic target functions were used to measure the performance of an individual and assign fitness: sine (fig. 4.3.1) and step (fig. 4.3.3), with the first requiring more smooth changes of protein levels and the latter a Boolean like pattern. While the induced cell's desired output would be in the the same phase as the input, ultimately the other cell had to produce the inverse of the input – which is equivalent to shifting the input's phase by one half ($\frac{wl}{2}$). Fitness was measured as the deviation from the corresponding desired output, i.e. the smaller the value, the better adapted the GRN.

Letting $c_{i_0}^t$ denote the (unbound) amount of the induced GRN's output protein i_0 and d_p^t the desired output in phase p relative to that of the input at time t the deviation is simply calculated as: $\sum_{t=1}^{L} |c_{i_0}^t - d_{0.0}^t|$ and again for the second cell, only with $d_{0.5}^t$, i.e. $\sum_{t=1}^{L} |c_{i_0}^t - d_{0.5}^t|$ – afterwards both values were added up and divided by 2.

However, in one condition individuals were not expected to fully differentiate immediately, i.e. from the first generation. Instead, the environment (i.e. the fitness function) became *gradually* harder by increasing the relative shift in wavelength (i.e. difference between the two fitness functions) little by little from 0 every 25 generations to $wl/2$ time steps (a phase shift of 0.5 respectively), when full differentiation was required. Writing g for the current generation, $\sum_{t=1}^{L} |c_{i_0}^t - d_{p^*}^t|$ with $p^* = \min(\frac{g}{\lceil 25 \rceil}, \frac{wl}{2})/wl$, was wanted for the second cell – so full differentiation was only required after 250 generations with the wavelength $wl = 20$ used in all experiments reported here.

The lifetime L of every individual was set to 400 time steps. A randomly-generated initial GRN could typically achieve a deviation of approximately 200 over this time. Finally, this value is used to transform the deviation to a standard 0 to 100 performance scale: $(200 - D)/2$, so zero deviation would result in a perfect performance value of 100.

4.4.2   Experimental Set-Up I: Impact of Selection Regime on Evolvability

Overall, 8 evolutionary scenarios were tested (two desired output types times four environmental stimulus input functions) and each scenario was run ten times. The whole experiment was run with immediate and with gradually increasing environmental differentiation pressure in a condition of perfect matching and with a fixed number of genes. To make sure that the task is not trivial the space of two types of randomly created GRNs was searched. The same number of random GRNs as would be generated during a normal experiment (10 times $125,000$ per type) were evaluated.

4.4.3   Results I: Impact of Selection Regime on Evolvability

For the two random scenarios average as well as best performances stayed well below evolved performances, cf. 4.3. In the evolutionary scenarios most runs successfully produced well adapted individuals that had evolved a kind of switch, allowing
them to behave very differently when an inducing stimulus was present. Not very
surprisingly, the more sparse the input was the harder it was to reduce the deviation
from the desired output wave. Unfortunately, in some runs no GRN in the final population could be considered to have an acceptable performance level. For the gradual
condition, this failure happened only once (in all 80 runs), and the superiority of
this condition can also be seen from table 4.3. It seems that an evolutionary environment that gradually becomes harder facilitates differentiation, and the smaller
standard deviations suggest an increase in robustness indicating higher evolvability.
This is also reflected by the finding that for most experiments the best/worst runs
are closer together when the lineage's environment changed slowly; for an example
see fig. 4.12.

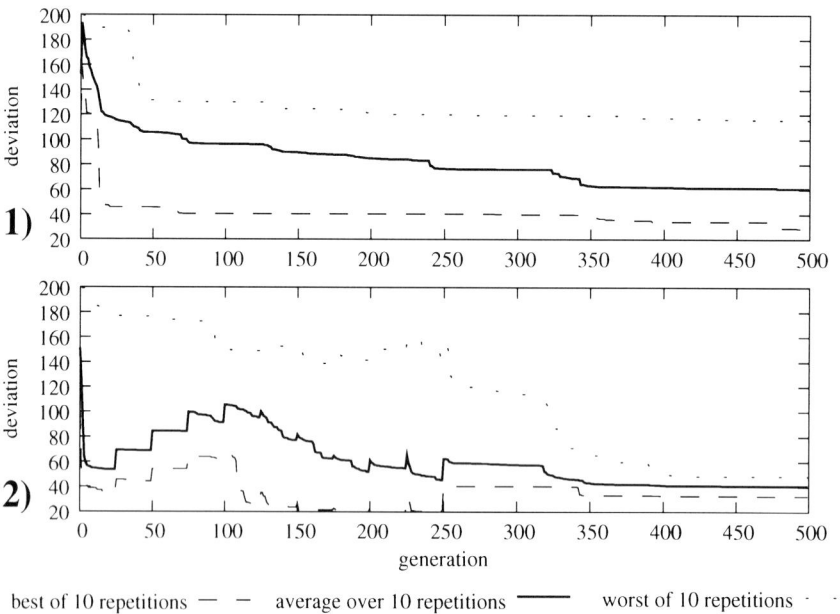

best of 10 repetitions — —   average over 10 repetitions ⎯⎯⎯   worst of 10 repetitions · ·

Fig. 4.12 Exemplary evolutionary runs showing the best individual per generation (average
over 10 runs); 1) with full differentiation pressure, 2) with gradually increasing differentiation
pressure. For most experiments the best/worst runs were found to be closer together when the
lineage's environment changed slowly.

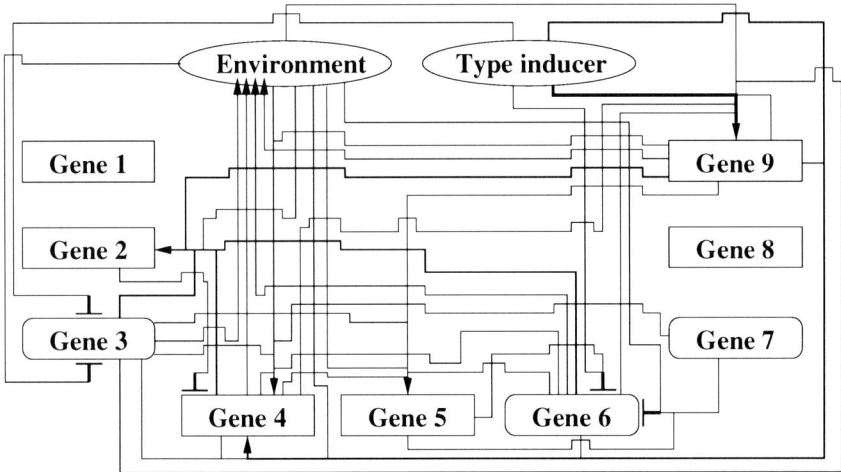

Fig. 4.13 Regulatory interaction diagram of an evolved 9-gene GRN. Boxes denote genes (rounded corners indicating "default on" ones with the others being "default off"), connections ending in an arrow are for activatory influences and the T-like endings depict inhibitory ones. The bolder the connections the more binding sites the receiving gene has for the corresponding protein, resulting in a bigger share of the protein binding.

4.4.3.1   Evolved Dynamics

In all the best evolved GRNs the use of new regulatory logic was found, with several binding sites bundled to a cis-module as described above. Typically, the protein level being influenced by the type inducer – which might be considered as the output of a "master control gene" or an environmental stimulus – had a very prominent position in well adapted individuals. For example the one shown in fig. 4.13 participates in the regulation of 4 out of 7 functional genes. In experiments (not shown in table 4.3) where the number of binding sites was limited to stay at one (no unequal crossover), evolvability was far worse – evolved performance was almost 20 percent lower on average compared to GRNs evolved without this restriction.

Finally, figures 4.14 and 4.15 illustrate protein amount and gene activity for an examplary individual, which was the best of its run in the scenario: perfect matching, sine input, sine output desired, gradually increasing differentiation pressure. The figures show its dynamics, with the lower matrices each corresponding to the second, "mirror behaviour desired", cell.

4.4.4   Experimental Set-Up II: Impact of Regulatory Mechanism on Evolvability

As in the first experimental set-up, 8 evolutionary scenarios were tested and each scenario was run ten times. As the gradually increasing differentiation pressure had

Table 4.3 Immediate and gradual differentiation pressure experiment outcomes, with the leftmost column depicting the environmental stimuli used and the topmost row the desired output behaviour for every run. The number of genes was fixed to 9 and perfect protein matching was used. Additionally performances of two kinds of random GRNs are given (one/two binding sites per gene). Data cells show the best final deviation averaged over 10 runs with 500 generations times 250 individuals each, ± the respective standard deviation.

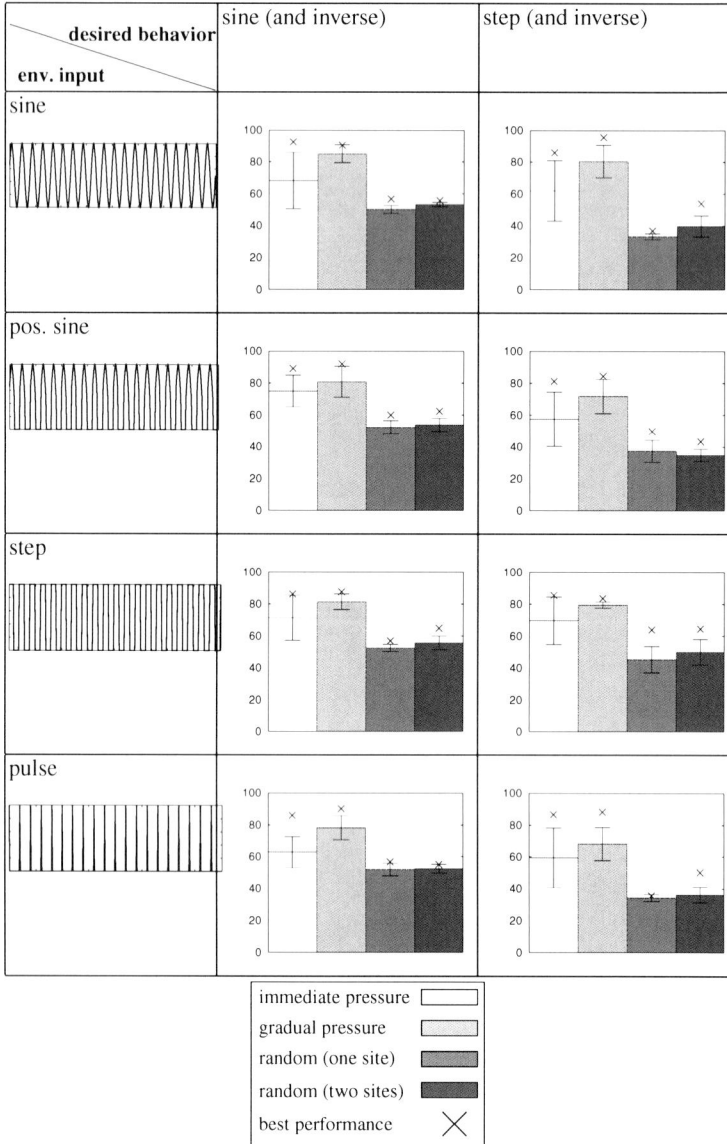

desired behavior / env. input	sine (and inverse)	step (and inverse)
sine		
pos. sine		
step		
pulse		

Legend:
- immediate pressure
- gradual pressure
- random (one site)
- random (two sites)
- best performance ×

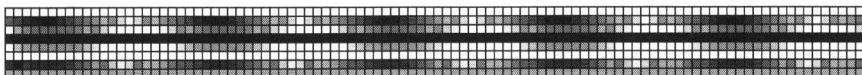

Fig. 4.14 In these matrices the *8 protein amounts* of the GRN from fig. 4.13 over 100 time steps are depicted. Note that row 2 reflects the input protein level while row 7 corresponds to the GRN's output. In the lower matrix lack of activity in row 4 induces inversion of the input stimulus.

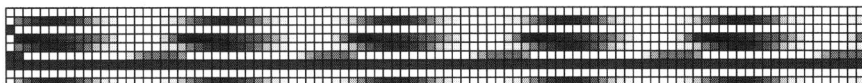

Fig. 4.15 As in fig. 4.14, but here the *output activity of each of the 9 genes* is shown. Every row corresponds to one gene's protein output; where darker means more output. One can clearly see the distinct activation patterns. Note that genes 1 and 8 generate no output ever, cf. fig. 4.13.

proven superior only this selection regime was used, and the number of genes was fixed again. The whole experiment was repeated with two different regulatory mechanisms: Perfect matching and smooth matching plus specificity factors[7].

4.4.5   Results II: Impact of Regulatory Mechanism on Evolvability

Using smooth matching and specificity factors did not immediately, in the 3 bits / 8 protein types setting, lead to an increase in performance. The reason was probably that the small number of proteins created too many interferences for fine grained regulation[8], as the picture changed when the number of proteins and thereby the search space was increased. While results for the perfect matching condition became worse with every bit added, the average performance for smooth matching

---

[7] Results for experiments with smooth matching but without specificity factors can be found online at http://panmental.de/GRNdiff.

[8] Interference can be useful in evolutionary search, potentially allowing the algorithm to locate "promising circuits". These however need to be fine tuned later in evolution to achieve the highest fitness values, but in small protein search spaces it can be impossible to avoid interference.

Table 4.4 Perfect vs. smooth (plus specificity factors) protein matching differentiation experiment outcomes, with the leftmost column depicting the environmental stimuli used and the topmost row the desired output behaviour for every run. The number of genes was fixed to 9 and gradual differentiation pressure was used. Data cells show the best final deviation for runs varying the number of bits used to encode a protein. All values are averaged over 10 runs with 500 generations times 250 individuals each, ± the respective standard deviation.

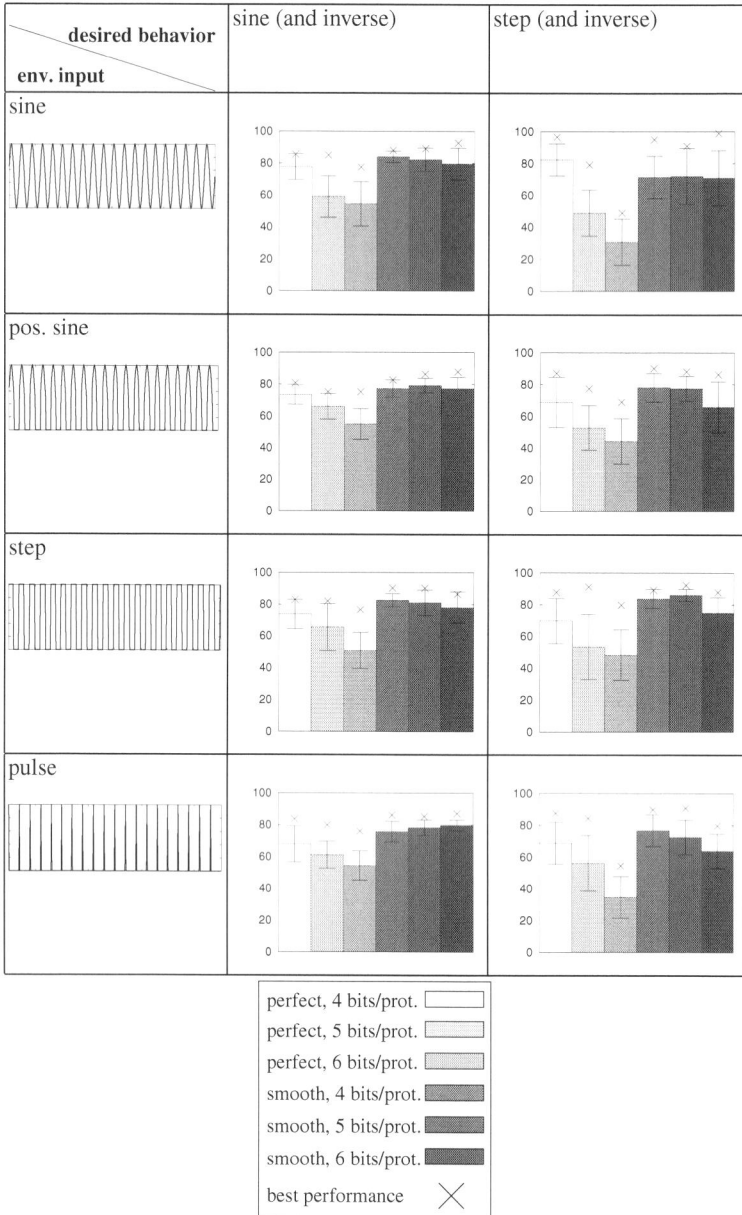

desired behavior / env. input	sine (and inverse)	step (and inverse)
sine		
pos. sine		
step		
pulse		

perfect, 4 bits/prot.
perfect, 5 bits/prot.
perfect, 6 bits/prot.
smooth, 4 bits/prot.
smooth, 5 bits/prot.
smooth, 6 bits/prot.
best performance ✕

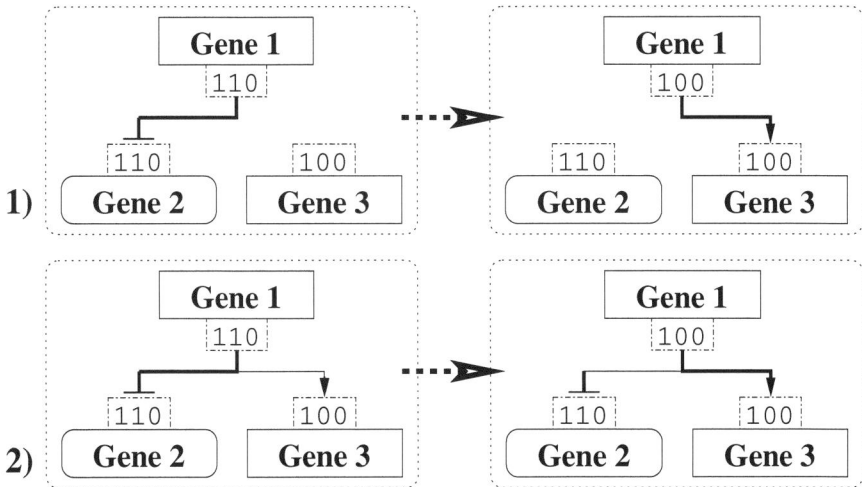

Fig. 4.16 Schematic drawing of the change of the network structure after one single bit mutation occurs, for 1) the perfect matching and 2) the smooth matching condition. Bolder lines represent stronger regulating influences. Note however that specificity factors can complicate this picture by dynamically changing affinities.

deteriorated very little with increasing bit numbers, cf. table 4.4[9]. Almost all evolved GRNs were found to make use of SFs during lifetime, although there was not a clear trend towards less or more use of it over evolutionary time.

4.4.6 Discussion

The xBioSys GRN model with two "layers" of regulation (synergistic effects between TFs within and additive effects between cis-modules) is clearly able to evolve functional differentiation. When exposed to a selection regime that selected for full behavioural differentiation from the first generation average performance of evolved GRNs was worse compared to a gradual increase in environmental pressure for differentiation. Also the variance of the evolved GRNs' performance was smaller in the latter case suggesting an increase in evolutionary robustness. The lineage's evolutionary history seems to be very important in determining the probability that a switch for differentiation between two behaviours can be found.

In setup II two protein matching mechanisms were studied: Simple template matching (also referred to as perfect matching) and smooth matching complemented with specificity factors. For smooth matching a regulatory protein is only more likely to bind binding sites it exactly matches but with some probability it can also bind to sites of a different type, thus making the mutation operator less

---

[9] Note that equal numbers of protein bits are not directly comparable as for the smooth matching condition one protein bit is used as a "specificity factor" flag.

destructive. The more different site and protein type the less affinity for binding there is. The probability distribution that determines how strong the affinity towards perfect matches is can be changed by specificity factors during a GRN's "lifetime" thereby adding a level of regulation. The experiments show that the use of smooth matching and specificity factors, compared to a static binding mechanism, can increase performance of evolutionary algorithms – especially for larger search spaces.

However, the use of smooth matching alone will also increase *pleiotropy* (one gene influencing multiple phenotypic traits). A pleiotropic gene might cause problems when genetic changes to it improve one trait while doing the opposite to another trait. Here specificity factors might be advantageous: They can modify the level of the smoothness of the distribution of transcription factor proteins binding to potential sites but their activity also influences how genetic variability will change network dynamics. Depending on lifetime specificity factor levels, a mutation could have any effect between the "perfect matching" condition and almost no effect at all for very smooth distributions. This difference is schematically depicted in fig. 4.16.

As [Altenberg(1995)] noted: "Genes are selected on for their organismal fitness effects but modify the variational properties of the genome as a systematic side effect". These effects and whether there is much interference between functionalities depends on the possible forms the distribution curve (see fig. 3.8) can take, which might be an interesting subject for further research. Also it would be interesting to see if the presented positive results still hold for more complex differentiation scenarios or other network formalism variants.

Chapter 5
Topological Network Analysis

Understanding GRN dynamics is a hard task and so methods for breaking down their complexity in more easily analysable parts have been proposed. Very influential has been the so-called "motif analysis", a general analysis method for all kinds of networks. This might be due to its simplicity, as it only searches for patterns (subgraphs) in the static, time-independent, connectivity structure of networks. Network *motifs* "are those patterns for which the probability P of appearing in a randomised network an equal or greater number of times than in the real network is lower than a cutoff value" [Milo et al(2002)Milo, Shen-Orr, Itzkovitz, Kashtan, Chklovskii, and Alon]. [Alon(2006)] and co-workers have developed the approach and used it to analyse the patterns networks are made up of, from the gene regulation network of *E. coli* to the world wide web. They have also shown what functions some overrepresented motifs might serve by analysing their range of dynamics exhibited in isolation. [Conant and Wagner(2003)] have suggested that network motifs were independently selected for particular functionality in a converging manner.

As pointed out earlier, differentiation is the basis for modularity and development. Accordingly it would be useful to find one or a few small circuits responsible for differentiation. This would make it easier to understand what GRNs are doing and to construct GRNs with particular functionality – by hand or by biasing an evolutionary algorithm, possibly also increasing evolvability.

5.1 Do Motifs Reflect Evolved Function?

The main research questions motivating this analysis are: Are there significant structural patterns that arise in the course of evolution in GRNs necessary for controlling the realisation of particular functionality? Are some motifs more prevalent than others in evolution for particular functions? How unique are the networks that realise particular functionalities and how robust are they?

These questions cannot be adequately answered on the basis of the current data available for natural GRNs; for this type of statistical analysis, many networks with

J.F. Knabe: Computational GRNs: Evolvable, Self-organizing Systems, SCI 428, pp. 71–81.
springerlink.com

a particular functionality must be compared and currently data are possibly incomplete and only available for networks in, at most, a few organisms. In addition, observing GRN interactions and analysing metabolism *in vivo* is very difficult and techniques are still in their infancy.

The "basic idea is that patterns that occur in the real network much more often than in randomised networks must have been preserved over evolutionary timescales against mutations that randomly change edges" [Alon(2006)]. The implicit assumption in this statement is that selection protects "useful structural modules" against a background randomisation process which randomly changes edges.

However, one has to keep in mind that mutation is not the only variability operator; in nature (and many models) recombination and duplication of genome constituents may take place, likely influencing network structure. Even if it were, it is not clear that in nature mutation acts like random rewiring (i.e. that a node is equally likely to connect with each other node) as used for example by [Alon(2006)]. The use of such a rewiring randomisation (without observing constraints of the natural process, as for example spatially close nodes might be more likely to be connected) procedure applied to a given network of interest might itself bias results. Luckily we do not have to work with the data limited to just one or two biological networks, due to the difficulty of determining GRN structure in natural systems, but have a fully controllable model at hand. So we are able to create even better random networks. "Even better" as the random network set can be generated using the same process, i.e. using the exact same variability operators on them for the same number of generations as the evolved networks, only without selecting for any function, thereby avoiding possible randomisation biases as discussed by [Artzy-Randrup et al(2004)Artzy-Randrup, Fleishman, Ben-Tal, and Stone]. This is termed *random evolution* in the following, i.e. standard variability operators but no selective pressure over the same number of generations.

From every such run only one randomly created network was used for analysis to avoid sampling biases and to make sure that network change over evolutionary time was independent. Note that all populations (whether for random evolution or evolution with selection) were generated from the same single random GRN, and from this GRN the same initial population of GRNs was created – using standard variability operators operators as described below. Only then, from this same starting population, each evolutionary run was carried out using a unique random number generator seed.

For analysis, all possible n-node sub-graphs of a network are enumerated, their connection matrix brought into a canonical form and occurrences of each unique pattern counted. The canonical form is defined here as being the smallest base 2 interpretation of the binary matrix entries taken in the usual order under all permutations of the node indices, see figure 5.1.1 for an example. To keep analysis concise only results for sub-graphs of three nodes are reported as these are most commonly found in the literature[1].

---

[1] Additional results for four-node motifs and Java code are available at
http://panmental.de/motifs/

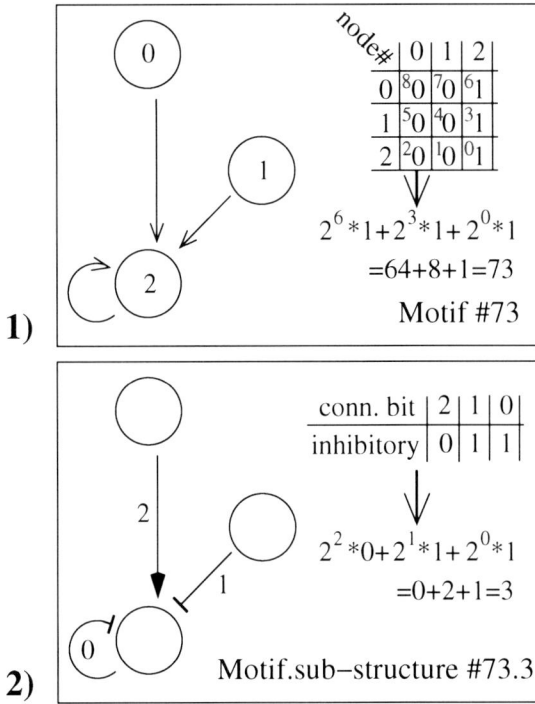

Fig. 5.1 Coding of motif and sub-structure number examples. 1) Motif to decimal integer; Of all possible node numberings the one resulting in the smallest motif number is used. 2) Activatory/inhibitory sub-structure coding – An arrow indicates an activatory influence while T-like endings depict inhibitory connections. Note that the connections 2 and 1 are interchangeable (an exchange would result in bit pattern 101, i.e. #73.5), but again always the smallest option is taken, #73.3 here, by convention.

In the literature the pattern count within a single known network from biology is generally compared against the pattern distribution (average number and standard deviations) from a larger number of randomised networks. As an arbitrary number of networks evolved for function can be produced here, additionally many against many can be compared. Furthermore a more fine-grained analysis, discerning sub-structures of motifs found, is introduced. Every connection in the GRNs can be either activatory or inhibitory, resulting in 2 to the power of the number of connections possible sub-structures per motif[2]. As connections have an order given by the motif's canonical form, writing 1s for inhibitory and 0s for activatory connections, one arrives at the binary encoding of sub-structures as shown in fig. 5.1.2.

---

[2] Actually "2 to the power of the number of connections" is only an upper bound, as two sub-structures can be identical apart from node numbering, to make the enumeration mapping unique only the smallest possibility is taken, cf. fig. 5.1.2.

5.1.1 Results

Population sub-graph patterns for the original, differentiation and random evolution conditions are compared. Every population consists of 80 individuals, each of them the best-performing individual from generation 1000 (or 500 where mentioned) of one independent run, i.e. there were 80 independent runs per condition[3].

All analysed networks had 9 gene nodes, meaning that $\binom{9}{3} = 84$ three-node patterns had to be counted per net. The last paragraph of this section makes an exception, as there experiments were run with 3 gene nodes, exactly one pattern – Analysability at the expense of performance.

There are more three-node patterns possible than appear in the plots, but for conciseness only those which had an average number of occurrences within the GRNs of a condition exceeding 0.5 are shown, i.e. the motif needs to be in at least every second network on average. See fig. 5.5 for all such patterns found in this study, which allows mapping from the unique numbers in the distribution plots to sub-graph patterns. Sub-graph patterns vary greatly between networks in each condition (cf. standard deviations in fig. 5.2). The differences in pattern distribution between the conditions with selection pressure are not strongly significant. Even for those motifs which are relatively frequent there are always some networks in which the corresponding motif does not occur at all. Unfortunately the added complexity of the differentiation task is not apparent from this analysis.

When comparing the two functional conditions to the randomly evolved condition some bigger differences show up. It should be mentioned that these differences become more marked with increasing generation number although fitness does not increase much after generation 500. This is not too surprising as, once the evolutionary algorithm has found a good solution, most changes to the GRN will be detrimental – while random evolution is totally unconstrained.

Making analysis more fine-grained by discerning different sub-structures of motifs also did not bring strong over-representation of a particular connection pattern to light. Considering the example of the distribution of activatory/inhibitory sub-structures of the most common motif #73 after 500 and 1000 generations (see fig. 5.3) strong peaks for graphs where all three connections are either activatory or inhibitory can be seen (#73.0 and #73.7 make up for more than 50% of occurrences of motif #73 in fig. 5.3). However, similar peaks are also found in the randomly evolved population. Accordingly selection pressures can not be the reason for this peak, making it likely that this is an artifact due to the composition of the initial population and variability operators. Only the breakdown of motif #73 into sub-structures is shown, as this pattern occurred most often and thus was least susceptible to random fluctuations. Data for the other motifs can be found online at http://panmental.de/motifs/

When comparing sub-graph patterns of individual (instead of groups of) evolved networks against the distributions of many random ones there was always at least

---

[3] N.B.: The word population is not used to refer to the population of a single evolutionary run here.

Motif distribution after generation 500 and 1000

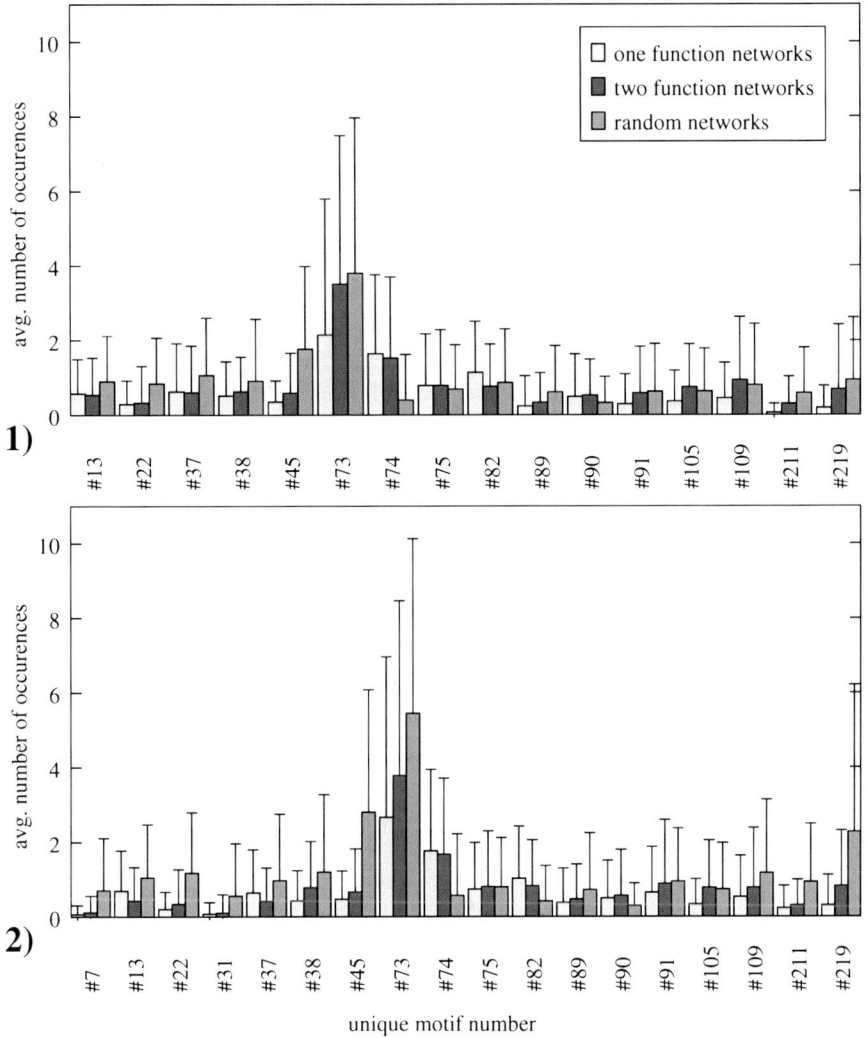

Fig. 5.2 Population average network 3-node motif distributions (9 gene networks), after 1) 500 and 2) 1000 generations. Three groups ($N = 80$ each) are compared: Original evolutionary condition for one task (light grey), evolved for differentiation (dark grey) and randomly evolved (medium grey). Only sub-graph patterns that have more than the cut-off value of 0.5 occurrences per GRN on average in at least one of the groups are shown. The unique motif numbers can be mapped onto graphs using the table in fig. 5.5.

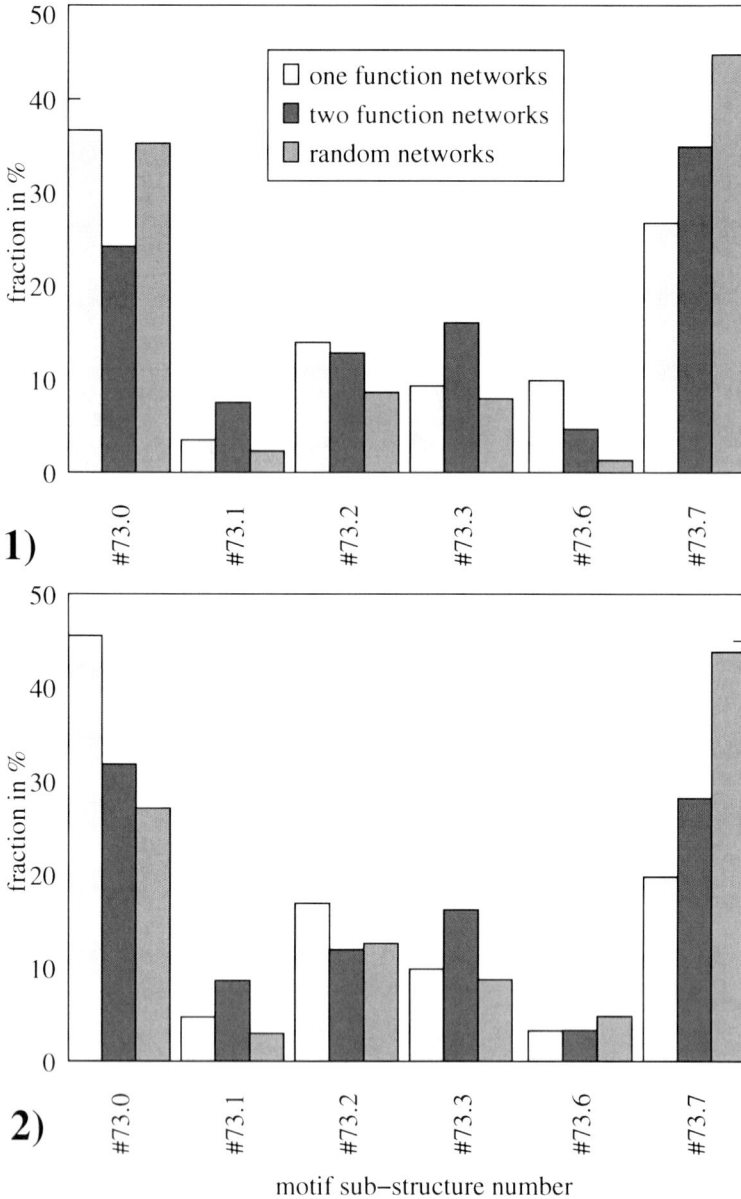

Fig. 5.3 Activatory/inhibitory sub-structure distribution on the example of motif #73 (9 gene networks). 1) after 500 generations, 2) after 1000 generations. Note that numbers #73.4 and #73.5 are not shown as they are identical to #73.2 and #73.3 apart from node numbering. See fig. 5.1 for motif #73 and text for details.

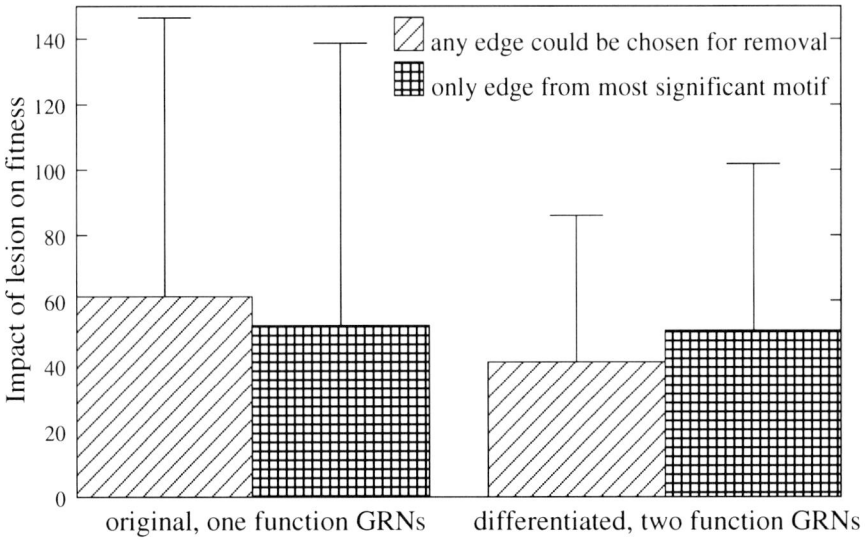

Fig. 5.4 Impact of lesions (9 gene networks): Average reduction in fitness when one edge in a network is removed for the original (left) and differentiated GRNs (right). In each pair, the diagonally striped bar shows the impact when any edge can be chosen while for the grid patterned bar an edge from the nets most significant network motif was removed. On average the differences are very small and the standard deviations huge.

one significantly frequent motif found. In order to check if the most significant motifs – mostly different between networks – are functionally very important to their GRN, a lesioning experiment is performed: From every best evolved network one binding site of a gene was taken away, either a) randomly any site or b) a random binding site from a sub-graph of the most significant motif of this GRN. Results of running these disrupted networks and measuring their performance drop have huge standard deviations but there are on average no big differences between a) and b), cf. fig. 5.4. Fitness (with a random GRN achieving around 200 as a reference) became on average worse by 61.2 ± 85.31 resp. 52.39 ± 86.27 for a) and b) in the original condition and 41.17 ± 44.83 resp. 50.75 ± 51.03 in the differentiation condition, revealing no significant robustness differences between lesioning a motif as opposed to a random location in either the original or differentiated conditions.

To check for the effect that *redundancy* and *entanglement* in the overall network has on motifs, the whole set of experiments was run again, but this time every GRN was restricted to only three gene nodes, thereby putting a lot of pressure on minimal design. Entanglement was given as a possible explanation of the variety of functions motifs can assume by [Mazurie et al(2005)Mazurie, Bottani, and Vergassola] and [Isalan et al(2008)]. With the smaller number of

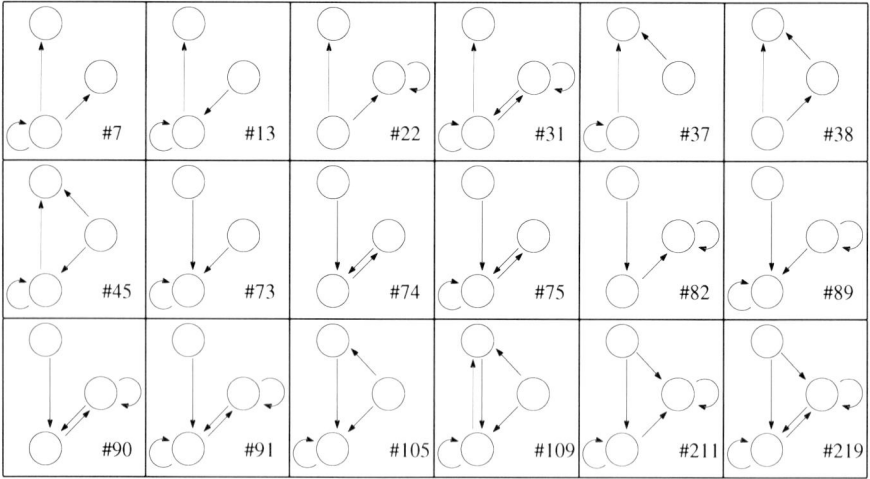

Fig. 5.5 The 3-node sub-graph network patterns occurring more often than 0.5 times per individual within evolved 9-node GRN populations on average. (The well-known feed-forward loop has number 38.)

gene nodes overall performance was worse compared to the 9-node networks used before, but the best evolved networks could still perform their respective functions acceptably. The differences between the best evolved GRNs from the best and worst runs however were bigger than before and, generally, it took more generations to find good solutions, so a higher number of gene nodes seems to facilitate more robust functional evolution. For the motif analysis it is important to note that now one GRN corresponds to exactly one 3-node motif pattern, and as motif occurrences were still widely distributed the cutoff value had to be lowered to 0.025 – i.e. there needed to be two networks with the same topology in (at least) one group of 80 to be shown in the plot. The surprising finding from the plot in fig. 5.6 is the over-representation of motif #73 as compared to the random condition. In both cases of functional evolution this motif represents the topology of about 30% of the best evolved networks, and remarkably most motifs found more than a few times in selection conditions contain the structure of #73 (*plus some other connections on top*, apart from #45, #82 and #90 as well as some that are not shown because they were not found twice). So apparently this configuration has some functional advantage, but unfortunately this seems not to be limited to one of the evolutionary conditions only.

When looking at motif sub-structures, patterns with two or three negative connections are most commonly found (cf. fig. 5.6) which is surprising considering the large fraction ($> 25\%$) of activatory-only configurations (#73.0, see fig. 5.3) found in the 9-node networks.

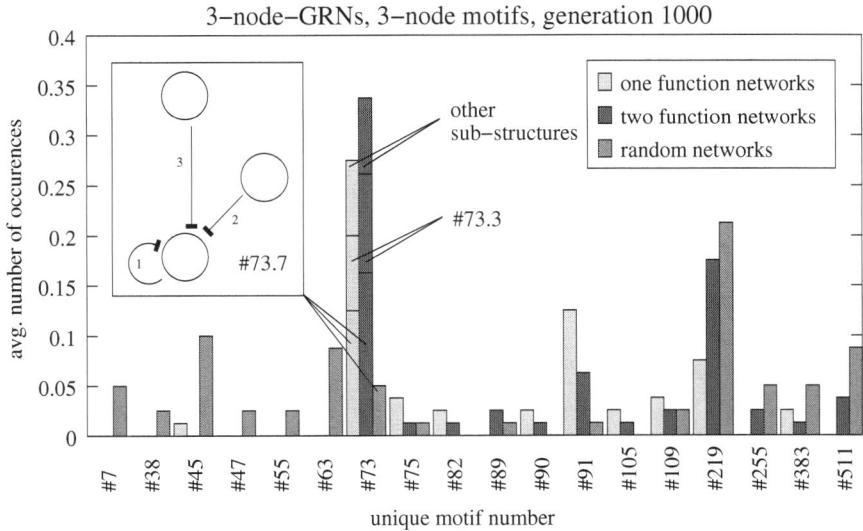

Fig. 5.6 3-node network motif data (3 gene networks). Three groups ($N = 80$ each) are compared: Original evolutionary condition for one task (light grey), evolved for differentiation (dark grey) and randomly evolved (medium grey). Unlike the distributions in fig. 5.2, these data are from GRNs restricted to only 3 nodes, so every network corresponds to exactly one 3-node motif, all motifs that occurred at least twice (0.025) in one condition are shown. For motif #73 additionally the prevalent sub-structures are indicated. Refer to fig. 5.1.2) for a depiction of sub-structure #73.3. See text for details.

5.2 Literature Review on Motif Analysis

Some other published research is also critical of the motif analysis method. [Mazurie et al(2005)Mazurie, Bottani, and Vergassola] compare available data from *S. cerevisiae* to four species that belong to the same class of hemiascomycetes, finding that "network motifs undergo no special evolutionary pressure as compared to a generic interaction pattern". Investigating the function of motifs they conclude that in the limited number of cases where enough information was available no specific functional role could be found. As a likely reason they suggest that entanglement with the rest of the network and the higher order context such as post-transcriptional regulation are more important in realising information processing capabilities.

[Artzy-Randrup et al(2004)Artzy-Randrup, Fleishman, Ben-Tal, and Stone] warn against drawing too far-reaching conclusions from motif analysis as one might be using an incomplete null model, i.e. randomisation of network data could introduce a bias in the analysis as it does not conserve spatial or other constraints to network topology. They show this by the example of a "Gaussian toy network" modelling the neural-connectivity map of *C. elegans*, where neurons are more likely to be connected the closer they are to each other. This model, without any action of selection, produces networks showing motif distributions similar to the ones found by [Milo et al(2002)Milo, Shen-Orr, Itzkovitz, Kashtan, Chklovskii, and Alon] in *C. elegans*.

[Kuo et al(2006)Kuo, Banzhaf, and Leier] also employ a GRN model and analyse static network topology. However the networks included in their systematic analysis were randomly created (focusing on whole genome duplication) without being exposed to evolution under selective pressure – and therefore were not comparable to that case and could not be lesioned to check for impact on function. The authors stress that methods of network construction, especially duplication and divergence events, could well explain motif distributions observed in natural regulatory networks. Similarly, [Ho(2007)], [van Noort et al(2004)van Noort, Snel, and Huynen] as well as [Cordero and Hogeweg(2006)] all devise mechanisms for the creation of networks structures. The generated topologies are very similar to those of real GRNs, without any selection for function or other sophisticated mechanisms needed.

Very recently [Isalan et al(2008)] were the first to systematically explore the effect of adding new links to the *Escheria coli*'s GRN *in vivo*. They find that network entanglement makes it often difficult to find functional modules: "Also, our results indicate that partition of a network into small modules (negative feedback, feedforward, and so on) could in some cases be misleading, as the behaviour of these modules is affected to a large extent by the rest of the network in which they are embedded."

5.3 Discussion

One might expect to find a switch motif to control differentiation when the requirement to differentiate was the only difference between the two evolutionary niches (target functions). That is, one might expect that motifs reflect evolved function. However the results show this view may be too naive – there was no convergence on the same single motif or a small set of switching motifs, and uniqueness of motifs was not observed. Instead a wide variety of network patterns and topologies was found.

Lesioning experiments showed that removing an edge from the most significantly represented motif in an evolved network has no significantly bigger impact on function than random removals. This suggests that particular motifs are not as important in the functioning of networks as might perhaps be expected.

As [Artzy-Randrup et al(2004)Artzy-Randrup, Fleishman, Ben-Tal, and Stone] have shown, randomisation of network data can introduce a bias in the analysis as it does not conserve spatial or other constraints to network topology. In this study a flawed null model or randomisation process was avoided by not randomising existing networks but instead generating them with the same process used for the functional networks. The only component absent in this random evolution process was selection for function.

Although many evolutionary runs were compared, the ability of the GRNs to produce similar dynamics employing a wide range of different topologies might still be to some degree due to the simplicity of the model and target functions used. Also in this model there is no cost for maintaining connections (regulating gene activity via protein production), however even in natural organisms it is not clear how costly transcription and thus the pressure for optimal design is.

Nevertheless the results warrant caution when topological measures like motif analysis are used to draw conclusions about functional properties. This point is reinforced by [Ho(2007), van Noort et al(2004)van Noort, Snel, and Huynen, Cordero and Hogeweg(2006), Kuo et al(2006)Kuo, Banzhaf, and Leier] who have shown that topologies strikingly similar to those of real GRNs can be created by simple mechanisms without selection for function.

[Mazurie et al(2005)Mazurie, Bottani, and Vergassola] have suggested higher order mechanisms such as post-transcriptional regulation or alternatively network entanglement as possible explanations for the range of different functions a motif can take on in different contexts as opposed to finding a specific, crucial role per motif. In the model used here however, all functionality is realised on transcriptional regulation level only, ruling the former explanation out. Entanglement in the network context seems to have more of an impact, as the experiments with GRNs restricted to 3 gene nodes (one GRN corresponding to exactly one 3-node motif) showed. This restriction put high pressure on minimal circuit design and here convergence was found to some degree (to the same structure for both settings with selection for function, average fitness worse compared to the 9 node networks). Thus, topological analysis might be hindered by the robust, redundant layout of networks, which is shaped by their evolutionary history.

Evolutionary convergence can be expected where there is a clear optimum much preferred over all alternatives and enough time and variation to find it. However for relatively simple problems with many solutions that perform similarly well, evolution will take what happens to be generated first, and these early random effects constrain what one ends up with [Gould and Lewontin(1979)]. This is the case in the 9-node GRN experiments as standard deviations of network pattern occurrences are very large even starting from the same initial populations, but much less so for 3-node networks. Perhaps in the latter case there is little room for "historical remains".

The connection matrix of GRNs might not to contain enough information to draw inferences on detailed functionality leading to deeper understanding. Among the properties not included in a connection matrix are grouping of inputs and non-linear protein interactions, inhibitory connections as well as connection strength and dynamic patterns, each of which can have decisive impact on function. For functional analysis it might be necessary to focus not so much on the structure of networks, but on dynamical ("metabolic") properties. Currently it is hard to get such data for biological systems – models could be at an advantage here.

Chapter 6
Development and Morphogenesis

In morphogenesis[1] dividing cells assemble into differentiated shapes, using decentralised control and self-organisation. The development of multicellular organisms from a single fertilised egg cell has fascinated humans at least since Aristotle's speculations more than 2000 years ago [Wolpert et al(2007)Wolpert, Jessell, Lawrence, Meyerowitz, Robertson, and Smith]. In the more recent past our understanding of how interacting genes direct developmental processes has greatly increased [West-Eberhard(2003), Gerhart and Kirschner(1997), Wolpert et al(2007)Wolpert, Jessell, Lawrence, Meyerowitz, Robertson, and Smith, Arthur(2000)], see earlier sections on evo-devo, 2.2, and its molecular basis, section 3.1.1. Cell differentiation, the inducing effects of intercellular signalling via "morphogens", changes in cell form like contraction, the self-organising properties of adhesion and cell sorting in animal morphogenesis [Glazier and Graner(1993)] are among the important principles better understood now. [Nehaniv(2005)] discusses GRNs as a potential computational paradigm with high evolvability. And although every cell is controlled by a Genetic Regulatory Network (GRN), the resulting multicellular dynamics are also strongly influenced by physical constraints.

6.1   The Physical Environment

Morphogenesis is inherently embodied in the physical environment. This characteristic makes "computing" more difficult due to stochasticity for example – it is not idealised and abstracted. However, embodied morphogenesis can make use of its physical realization, e.g. by using physical states and processes in place of explicit computational representations. Being situated in an environment also allows for adaptive phenotypic plasticity. In a standard work on Biophysics [Forgacs and Newman(2005)] write that it is an often neglected aspect that morphogenesis can exploit self-organising characteristics of the physical world.

---

[1] Also ontogenesis or ontogeny, an organisms history from birth – as opposed to phylogeny, the evolutionary history.

J.F. Knabe: Computational GRNs: Evolvable, Self-organizing Systems, SCI 428, pp. 83–100.
springerlink.com   © Springer-Verlag Berlin Heidelberg 2013

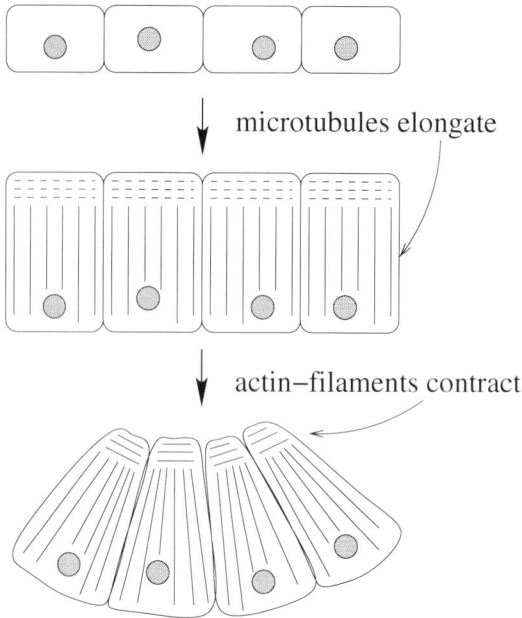

Fig. 6.1 Schematic bending of an epithelium through cell shape changes mediated by elongation of microtubules and contraction of actin filaments. After [Alberts et al(2002)Alberts, Johnson, Lewis, Raff, Roberts, and Walter].

6.1.1   *Spatiality*

Naturally, cells organise in physical space. The most important mechanisms to achieve particular embryonic forms are differential cell adhesion and motility as well as changes to individual cell shapes and sizes. Differential cell adhesion is achieved with selective "hook and loop" protein domains partially outside of the cell. The most famous of several groups of proteins acting in such a way are termed cadherins. Cells can bind to other cells expressing the right counterparts (the same molecule on an adjacent cell, helping tissues to form) but also to the extracellular matrix, which provides structural support to cells. Both mechanisms are also important for various other cell behaviours such as motility, where haptotaxis (along a gradient of adhesion sites) and chemotaxis (along a chemical gradient) have to be distinguished, among others.

Cells can also deform their inner structure or cytoskeleton, cf. fig. 6.1. In a highly dynamic but controlled process microtubules are continuously broken down and built up to achieve particular cell shapes. Of particular importance during cell division is the spindle apparatus, as it organises the split along a preferred axis and separates the chromosomes into the daughter cells. Furthermore, the dynamic network of actin filaments is responsible for cell protrusions, cell deformation and the final separation of the two daughter cells.

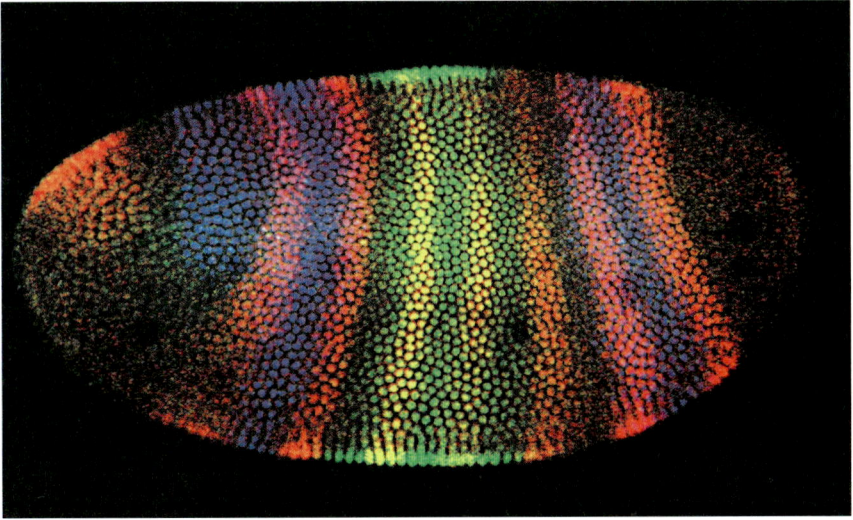

Fig. 6.2 Differential gene activation leading to pattern formation in *Drosophila* embryo. Each circle is a separate nucleus. Picture was created by attaching fluorescent proteins to certain transcription factors. (The "bio-marker" GFP (Green Fluorescent Protein) can be added without doing harm, making it relatively simple to see gene activation during - and after - morphogenesis.) From *Molecular Cell Biology*, 5/e by Harvey Lodish, et al. (c) 1986, 1990, 1995, 2000, 2004 by W. H. Freeman and Company. Used with permission.

6.1.2 Pattern Formation

During morphogenesis cells organise into astonishingly regular patterns of early cell differentiation which forms the basis of later body plan layout: tissues, organs and segmentation. Figure 6.2 shows such patterning in a Drosophila embryo. Higher level patterns can often be described with a small number of macroscopic variables. As already introduced in section 3.1.1, this emergent organisation is largely achieved by lower level communication mechanisms. Different modes of intercellular signalling are known. Some proteins transfer directly to particular neighbours via "gap junctions" and others remain on the cell surface of the originating cell so that all cells in direct contact are affected. In contrast to these targeted, local communication and induction mechanisms, here the focus will mostly be on diffusion.

Generally large molecules diffuse more slowly than small ones and bind to less specific receptors. Accordingly "medium range" diffusion is often not very specific. A combination of the varying distribution of diffusing molecules and the timing of encountering them induces cells to express the appropriate program [Wolpert et al(2007)Wolpert, Jessell, Lawrence, Meyerowitz, Robertson, and Smith].

Concentration of morphogen

1 2 3 4 5 6

Concentration of morphogen

thresholds

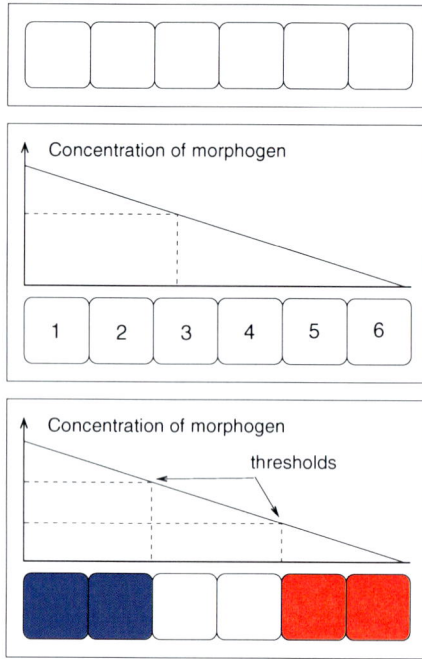

Fig. 6.3 French flag model. Over an array of initially identical cells the presence of a (stable) morphogen gradient is assumed. Each cell can sense the local concentration at its position. Based on threshold values the cells can now decide which cell type to assume. After [Wolpert et al(2007)Wolpert, Jessell, Lawrence, Meyerowitz, Robertson, and Smith, Wolpert(1996)].

6.2 Modelling GRNs: Development and Morphogenesis

One of the earliest researchers looking for a theoretical explanation of how cells in a developing embryo could establish their different roles was [Turing(1952)]. He proposed a general symmetry breaking mechanism via the setting up of chemical gradients with reaction diffusion systems, which is now known to be generally correct. Somewhat later, [Wolpert(1969)] came up with the very illustrative French flag model as an attempt to explain how morphogen gradients could give cells positional information as a general biological process. "Stem cells" placed along a given morphogen gradient would only have to read the morphogen concentration at their position and react to threshold values to decide whether they are in the blue, white or red part of the flag. The model is illustrated in figure 6.3. Note that this assumes the existence of a gradient but does not explain how such a gradient could be set up and maintained by the cells. [Jaeger and Reinitz(2006)] proposed a revised French flag model which to some degree takes the dynamic, feedback-driven nature of pattern formation into account. Even before that, [Meinhardt(1978)] designed a sophisticated reaction diffusion system morphogen gradient model for cell determination. Along similar lines, [Furusawa and Kaneko(1998)] models the emergence of differentiation and spatial patterns.

In contrast to these designed, theoretical models, here the focus will be on models that use evolutionary algorithms to evolve morphogenesis, an approach that has been variably termed "Artificial Development", "Artificial Ontogeny" and "Artificial Embryogeny". The models have in common some form of indirect, developmental genotype to phenotype mapping (not necessarily based on a genetic regulatory network). This adds a "physical dimension": however abstract the model is, the possible interactions between the developing parts influence the evolutionary algorithm. While still being far away from full physics and especially the thermodynamic laws, it has been argued that some embodiment is better than none and can provide evolvability advantages. [Kumar(2004)] for example found, that implicit developmental encoding[2] is more scalable compared to other encodings. A book edited by him and Peter Bentley gives a good overview of morphogenesis models at the time [Kumar and Bentley(2003)].

[Sims(1994)] however was the first to use a simulated world with somewhat realistic physics, including the simulation of gravity and friction. The evolutionary target for the individuals was to move as far as they could in their environment, among other tasks. No interaction with the environment occurred during morphogenesis and controller and body were encoded separately. The phenotype is constructed from a graph like structure, reminding of L-systems, from box-like primitive elements. Connections were relatively constrained (with few, pre-defined functions), for example joints could be of type rigid, revolute, spherical, twist, universal. The genome effectively stores the organism's structure, not its growth processes.

[Bongard and Pfeifer(2001)] and [Bongard(2002)] employ a developmental system to create impressive critters that also have to move in a physical world. Again, morphogenesis does not take place under physical constraints. Their model is relatively abstract as the units their creatures are composed of have a pre-specified cylindrical shape (different lengths possible) and complex internal elements like motoric joints. Symmetry breaking activation is induced by application of "morphogens" to the opposing ends of all new units.

[Eggenberger-Hotz(1997)] was one of the first to use discretised space. On a 3 dimensional grid of 30x30x30 units, with con-generic cells of a constant size of 1^3, he evolves 3D shapes. In an extension of this work, GRN-controlled joints between cells allow e.g. invagination [Eggenberger(2003)].

[Kumar(2004)] managed to evolve three to ten GRN controlled cells to form and maintain a concentration gradient, but cells did not have to interpret gradient concentration at their location.

[Miller(2003)] and [Miller(2004)] evolved cells to match a 12×9 French flag pattern, with a particular focus on self repair in the evolved individuals. There was a concept of cell division, but the possibility of overwriting of neighbouring cells during division might have been crucial for the outcomes.

This work was extended by [Federici(2004)], who evolved individuals to match less regular 9×6 patterns. An interesting analysis of why evolved

---

[2] Here "implicit" means the genome controls growth processes rather than storing structure.

developmental individuals are often also fault tolerant is given in follow-up work
by [Federici and Ziemke(2006)].

More recently, [Devert et al(2007)Devert, Bredeche, and Schoenauer] evolved
neural network controlled cells to match 32×32 pixel patterns, without division
(cells would be everywhere from the beginning). They focused on robustness and
an adaptable stop criterion – an individual was considered final when a stable state
was reached.

Crucially, in all these models cells were *homogeneous in size*, namely 1 pixel
(on a grid) each. Also, cell-cell communication was predefined as directed between
nearest neighbours (with fixed neighbours).

[Steiner et al(2006)Steiner, Olhofer, and Sendhoff] describe an interesting GRN
model with TF diffusion in a developmental shape optimisation setting. They do not
use cell adhesion or movement and rely on pre-defined TF gradients.

In recent work[3], [Joachimczak and Wrobel(2008)] present a 3D model of multicellular development. GRNs control parameters and TF diffusion, although some
TFs gradients are constantly present. The model supports spring physics to simulate
cell adhesion, and cells are balls that can differ in diameter.

Also recently, [Buck and Nehaniv(2008)] began to investigate the capacity of
multicellular GRNs to address NP-complete problems, in particular map colouring.

[Hogeweg(2000)] has used a model were cells also have spatial extent and can
move relative to each other. Cell types were under control of a Boolean GRN, where
the GRN's expression pattern was interpreted as one of the predefined cell types.
Differential gene expression was initiated by two pre-scheduled asymmetric cell
divisions (i.e. one gene would be on in one daughter cell and off in the other).
Communication was direct between nearest neighbour cells; a cell had the states
of two nodes from neighbouring cells as input to two of its own nodes. Unlike in
the other mentioned works, selection was not for particular cell arrangements, but
the number of exhibited cell types was used as fitness criterion – an evolutionary
algorithm would select for individuals that expressed many different types. The cell
type would determine some cell properties like adhesion (the possible adhesion values were pre-specified), so that certain multicellular forms could be observed as an
evolutionary byproduct.

6.3   Evolving Differentiated Multicellular Organisms

The discussed theoretical analyses and especially the "French flag" model have inspired the work described next. From a single cell, placed in the middle of a 60×40
pixel grid, a multicellular organism has to grow. The organism not only has to span
the grid but its cells also have to express different colours in different areas (blue,
white, red – from left to right). Unlike in the earlier experiments (see section 4.4)
and the related models mentioned above, differentiation is not induced externally

---

[3] The work was published in parallel with the experiments below at the eleventh Artificial
Life conference.

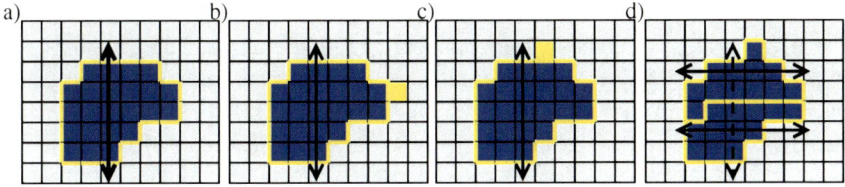

Fig. 6.4 Cell change example. The blue cell in a) has volume 23 and the orientation shown by the arrow. Assuming for this cell a target ratio of 1 between the length in arrow direction and the length in the direction orthogonal to the arrow. In b) the yellow pixel is tested for becoming part of the cell. However, as this would change the ratio from 6:5 to 7:5, the transaction is energetically unfavourable. The yellow pixel in c) brings us closer to the target length ratio so it is more likely to be accepted (target volume and other constraints permitting). In d) the cell reached the mitosis volume of 24 so it split evenly along an axis orthogonal to its orientation. The daughter cells inherit an orientation rotated by 90 degrees (which can be continuously modified by GRN dynamics).

at all. Instead, the following work focuses on the self-organised differentiation of cells. In this work there are no predefined communication channels; cells can only sense morphogen concentrations. Morphogens are excreted by cells and diffuse on the grid without preference for direction. As the focus here is on diffusion, other ways of cell-cell communication, like the direct contact of complementary proteins on the cell's surface and gap junctions, are not included here.

Cells can actively aim to adopt heterogeneous sizes and shapes – the importance of which for morphogenesis has been shown by other researchers, e.g. [Merks et al(2006)Merks, Brodsky, Goligorsky, Newman, and Glazier], [Zajac et al(2003)Zajac, Jones, and Glazier]. A cell's GRN also individually and independently controls other cell properties like morphogen secretion and cadherin expression. So there is no *a priori* notion of cell types.

Typically cells in fixed-neighbourhood-systems would not be allowed to divide if all neighbouring spots are filled or would overwrite a neighbour. In the present model a kind of pushing results when a cell grows or divides into neighbouring areas: it might overwrite a pixel of the neighbour if there is enough energetic pressure for growth, but this in turn will increase the other cell's growth pressure, making it more likely that it will get the pixel back from one of its neighbours and so on.

6.3.1 Spatial Model

The Cellular Potts Model (CPM) has been introduced by [Glazier and Graner(1993)], who developed it to simulate differential adhesion driven cell arrangement. Since then CPM has been used for a variety of cell-level modelling tasks, recently reviewed by [Merks and Glazier(2005)]. Although quite complex models have been realised with CPM, like the development of a cellular slime mould by [Marée and Hogeweg(2001)], the cell level CPM simulation has not been combined with a GRN controlling cell parameters individually, without predefined cell types, before.

CPM is a two level system: On the lower level there is a grid of pixels, while the higher level cells consist of any number of lower level pixels. Cells have properties like target volume, shape, etc. and deviations from these targets incur energy penalties. Every pixel on the other hand has an integer value assigned, designating it as belonging to the cell with that identifier or zero if it is part of the "medium" (empty space). The system changes by trying to copy over one pixel's value to a randomly chosen neighbour pixel (morphogen concentration values are not copied, instead they diffuse in a separate process described below). In the work described here, due to computational restrictions, neighbours are only the four next pixels, thus no direct interactions take place past this distance. In every time step ("Monte Carlo Step") one copy attempt is undertaken for every grid pixel on average. Copying is limited by energy constraints: Changing a pixel will change the energy properties of one (if either pixel is part of the medium) or two cells. The overall energy E is the weighted sum of all constraining properties. For example, let λ represent the resistance to compression, a_σ the current volume of cell σ and A_σ its target volume. In a model where cell volume is the only constraint, the energy would be calculated as:

$$E = \lambda \sum_\sigma (a_\sigma - A_\sigma)^2 \tag{6.1}$$

Other per cell energy constraints are added in exactly the same way. Copying is accepted with probability:

$$P = \begin{cases} e^{-(\Delta E - \delta)/kT} & \Delta E \geq -\delta \\ 1 & \Delta E < -\delta \end{cases} \tag{6.2}$$

In the work presented here the settings were offset $\delta = 0.0$, Boltzmann constant $k = 1.0$ and temperature $T = 2.0$. A two-dimensional, non-toroidal 60×40 pixel grid was used.

For the simulations use was made of a flexible open source CPM implementation called CompuCell3D (version 3.1.17), see [Glazier et al(2005)Glazier, Heiland, Swat, and Zaitlen, Cickovski et al(2007)] for an extensive introduction and source code. Through a modular plugin structure (where usually one plugin adds one energy constraint) it is easy to write extensions for the existing software (standard plugins include volume, surface, mitosis and connectivity constraints). The following focusses on modifications and plugins developed in the course of this work, while the reader is referred to the sources cited above for implementation and standard plugin details[4].

---

[4] For the experiments reported here the following plugins were used: Connectivity, VolumeLocalFlex (energy constraint for volume), ContactMultiCad (energy constraint for cadherin adhesion), Mitosis (cell division), FlexibleDiffusionSolverFE (diffusion of morphogens). The following plugins were specially developed for this work and are described in detail below: GRNcells (GRN controller), CellShapeControl (energy constraint giving the GRN some control over cell shape).

6.3.1.1   GRN Controller

A GRN is able to control the cell's volume via a protein level mapping to the ratio of the size required for mitosis. So a maximal protein level meant the cell would try to grow until mitosis took place while a zero protein level would initiate shrinking and usually lead to apoptosis within a few steps. The actual volume of the cell could however differ from the target size due to neighbouring cells for example. To take this and the fact that externally enforced size differences can affect the behaviour of biological cells [Folkman and Moscona(1978), Huang and Ingber(2000)] into account, the actual volume ratio served as input to the GRN, i.e. determined another protein's level.

   Two protein levels were used to determine a cell's colour. The continuous protein levels were interpreted as Boolean values for this (above 0.5 threshold: true, below threshold: false). The colour however does not correspond to cell type as in other models, where the type would determine cell parameters. All other parameters are controlled by the GRN independent of the colour chosen. Also the colour was not chosen once and for all but protein levels would be interpreted in every time step anew.

   The GRN could also control the expression of "cadherins", factors that influence adhesion to other cells expressing them, i.e. it could be to some degree energetically favourable for cells expressing cadherins to "stick together". In the experiments reported here there are three cadherins. Unlike other energy constraints that determine the probability of changes, adhesion is not calculated on a per cell, but on a per pixel basis. Let $\mathbf{x}, \mathbf{x}'$ be two pixels that are part of different cells (not the same cell, not medium), and $\tau_{\mathbf{x}}^{i}$ a function returning the expression level of cadherin i in the cell $\mathbf{x}$ belongs to. The adhesion energy was then calculated as:

$$\sum_{i-1}^{3}\sum_{j=1}^{3}\sum_{\mathbf{x},\mathbf{x}'} J^{ij}(\tau_{\mathbf{x}}^{i}, \tau_{\mathbf{x}'}^{j}) \qquad (6.3)$$

The adhesion strengths or characteristic bond energy J^{ij} between the three expressible cadherin proteins were pre-specified (see web page for table).

   An overview of all GRN inputs/outputs is given in table 6.1.

6.3.1.2   Diffusion

Morphogens diffuse and decay on the underlying grid (substrate) – so it does not immediately matter for diffusion whether a cell is present or not. A cell can however modify the concentration of a morphogen on the pixels it consists of. Unlike earlier models with directed cell-cell communication mechanisms, where a cell receives as input some output (often state) of its direct neighbours, this allows for (relatively undirected) long distance communication. On the other hand meaningful communication might be harder to evolve this way as it is less directed and morphogens do not correspond to other cell variables.

Table 6.1 Overview of all inputs and outputs a GRN has (proteins with other numbers, 14 and 15 here, have no pre-defined function). An output of the GRN that controls a property of the cell is referred to as parameter. Protein number is the integer value of the protein that is actually modified (for inputs) or read. See text for details.

type	property	protein number
parameter	colour 1	0
parameter	colour 2	1
output	morphogen 1	2
output	morphogen 2	3
parameter	shape	4
parameter	cadherin 1	5
parameter	cadherin 2	6
parameter	cadherin 3	7
parameter	size	8
parameter	preferred direction	10
input	morphogen 1	12
input	morphogen 2	13

The diffusion plugin "FlexibleDiffusionSolver" that was used is included in the CompuCell3D package and is explained (especially the numerical approximation) in its manual [Glazier et al(2005)Glazier, Heiland, Swat, and Zaitlen] so only a brief description is given here. The concentration c_i of morphogen i with diffusion constant d_i and decay constant k_i changes as:

$$\frac{\partial c_i}{\partial t} = d_i \nabla^2 c_i + k_i c_i + secretion_{ixy} \tag{6.4}$$

The term $secretion_{ixy}$ is the increase of morphogen i in pixel (not cell) xy. Here two morphogens were used, with the constants $d_1 = 0.2$, $k_1 = 0.009$ and $d_2 = 0.2$, $k_2 = 0.003$.

A cell receives the average morphogen concentration of the pixels it consists of as input (determining one protein level per morphogen). Two other protein levels determined the secretion of the corresponding morphogen, realised as increasing the average concentration of the cell's pixels.

6.3.1.3   Cell Shape Control

Apart from the plugin which manages GRN control over parameters determining cell dynamics, another major new module was developed: CellShapeControl. Earlier work successfully used a plugin to model cell elongation along the cell's longest axis [Merks et al(2006)Merks, Brodsky, Goligorsky, Newman, and Glazier, Zajac et al(2003)Zajac, Jones, and Glazier]. However, in these earlier works elongation followed the cell's longest axis, so orientation was due to initial random effects.

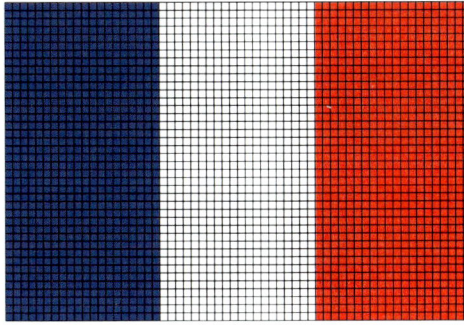

Fig. 6.5 Evolutionary target was a 60×40 pixel three striped pattern - the French flag

In this model the orientation is partially inherited and partially under the GRN's control (see mitosis section below for details). Furthermore, better results were found during evolution when not the length directly was under GRN control but the ratio of the length along the orientation axis to the length along the orthogonal axis, see fig. 6.4 a)-c). So in the experiments reported here, shape is loosely controlled by the GRN as the target ratio between the cell's length in its preferred direction vs. the orthogonal direction.

6.3.1.4   Mitosis

Mitosis was initiated automatically once the cell reached a volume of 24 pixels[5], implemented as equal split along the axis orthogonal to the cell's orientation, see panel d) in fig. 6.4. The heritable part of the cell's orientation was shifted by 90 degrees[6] and target volume set to half the parent's value in both daughter cells. Free protein levels were divided equally, while protein bound to binding sites stayed within one cell. The latter allows for autonomous differentiation, i.e. the daughter cells could "figure out" which cells is the original due to this.

6.3.1.5   Initial State

In the experiments reported here, "maternal factors" were not included, i.e. the first (and only) cell started without spatial extension at a size of 1 pixel, in the middle of the grid. All protein levels were set to zero for the GRN and no morphogen was present.

---

[5] Note that this is relatively unrealistic as at least the first cell divisions normally occur without cell growth.

[6] This is in analogy to the behaviour observed in nature, where centrosomes are known to rotate, so that the new spindle is at a right angle with the previous one [Wolpert et al(2007)Wolpert, Jessell, Lawrence, Meyerowitz, Robertson, and Smith, p. 264].

6.3.1.6   Fitness Evaluation

The lifetime of each individual was 200 time steps. An individual's fitness was defined simply as the match of the cell formation at the last time step compared with the 60×40 pixel target pattern comprising the entire spatial field, see fig. 6.5. Formally, the difference d between individual i's result pattern R^i and the target T, both of size $w \times h$, is:

$$d(R^i, T) = \frac{1}{wh} \sum_{x=0}^{w-1} \sum_{y=0}^{h-1} |sgn(R_{xy}^i - T_{xy})| \in [0,1] \qquad (6.5)$$

Accordingly, the correctness of the match or similarity s is simply $s(R^i, T) = 1 - d(R^i, T)$. Note that this measure does not specify the number, size or position of cells, only the lower level pixels are taken into consideration (making it nearly impossible for individuals to achieve a perfect match to the target, simply due to random fluctuations).

One refinement introduced was that the difference between the patterns would be multiplied by four minus the number of colours present in the final individual. An individual using all three possible colours would get a factor of one, while individuals using fewer colours would be penalised with a higher factor. Let c denote the number of colours (out of three possible ones) used by individual i, then the fitness is:

$$(4 - c) \times s(R^i, T) \qquad (6.6)$$

The aim of this was to "encourage" individuals early on in evolution to use all colours – without this measure runs were prone to get stuck in local maxima with huge single-colour individuals.

To improve robustness, every individual was run 10 times with different random seeds and the overall fitness calculated as the average of these repetitions.

6.3.2   Results

Seven out of the ten runs conducted with the particular configuration of plugins and parameters described here ended with similarities above 0.75. Recall, this is the average match over 10 repetitions with different random seeds. So a majority of evolutionary runs achieved final pattern matches of over 75 percent, with a typical course being shown in fig. 6.8. In particular, most individuals used morphogens to set up, and maintain, a gradient across the grid, as did the example in fig. 6.6. In the figure it becomes quite clear that morphogen diffuses on the underlying grid as you can see low levels of it in areas without cells.

For evolved individuals the cells would usually quickly divide and spread over the grid early on during an individual's lifetime while becoming more stable later – pattern similarities at time step 210 were very close to the originally measured fitness at time step 200.

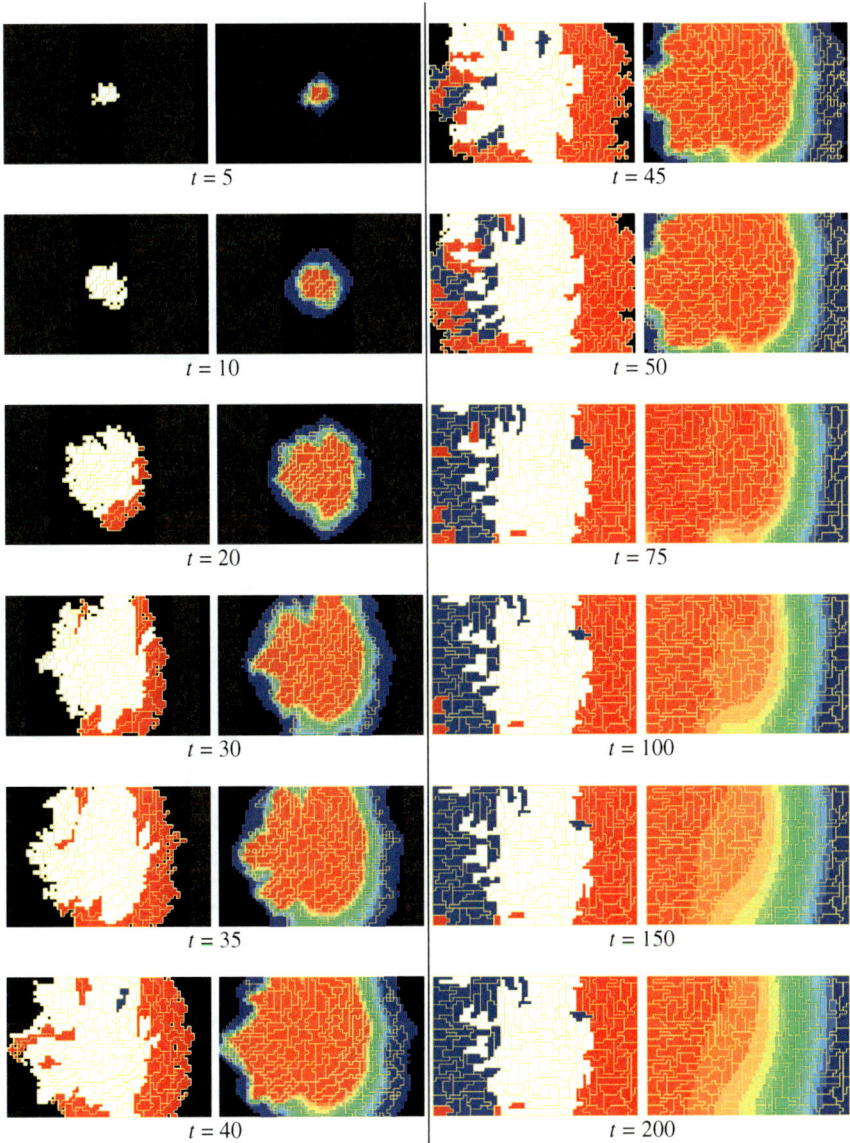

Fig. 6.6 Morphogenesis example. The left picture shows the colours of the cells while the right one shows adjusted morphogen level (red=high to blue=low), respectively. Cell outlines are given in yellow, t is the time step the snapshot was taken at. Similarity to the target French flag pattern at $t = 200$ is 87 percent, with 149 cells.

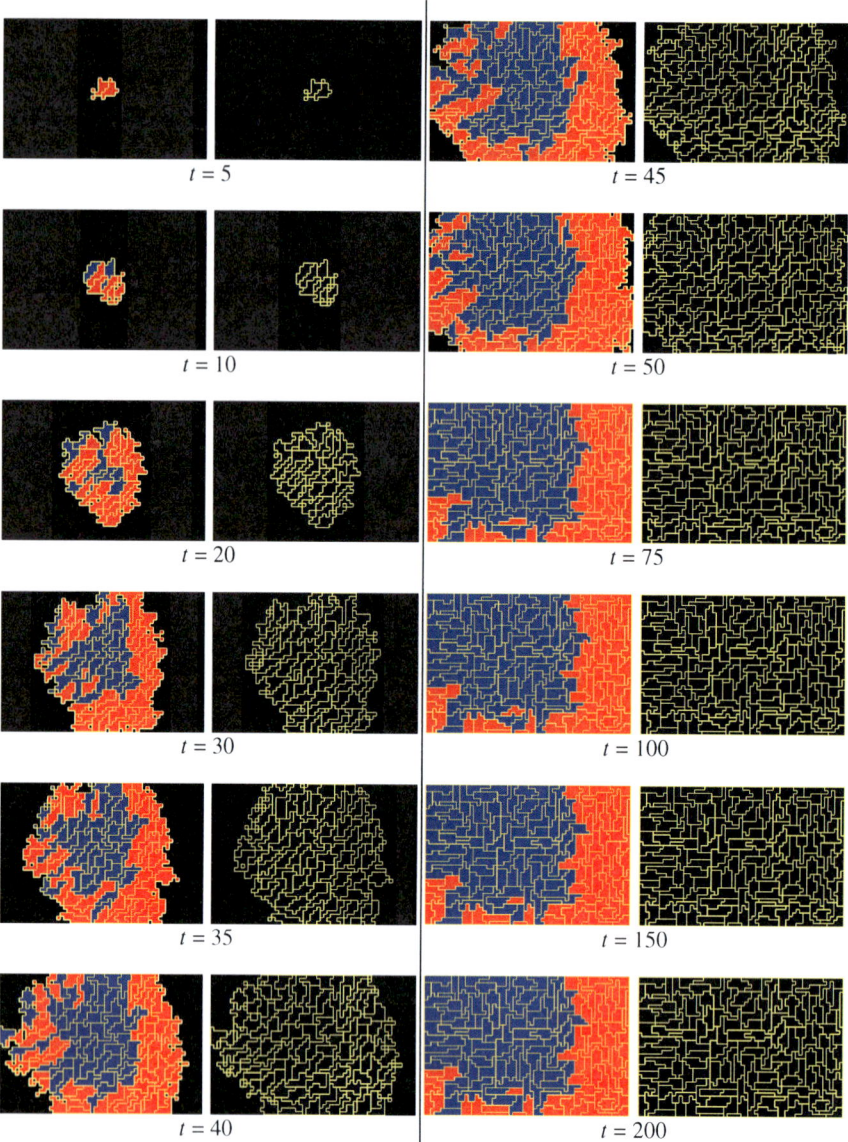

Fig. 6.7 Morphogenesis example. Same individual as in figure 6.6, but with morphogens disabled (black equals no concentration) to check for conditional differentiation. Clearly the cells require morphogen to fully differentiate.

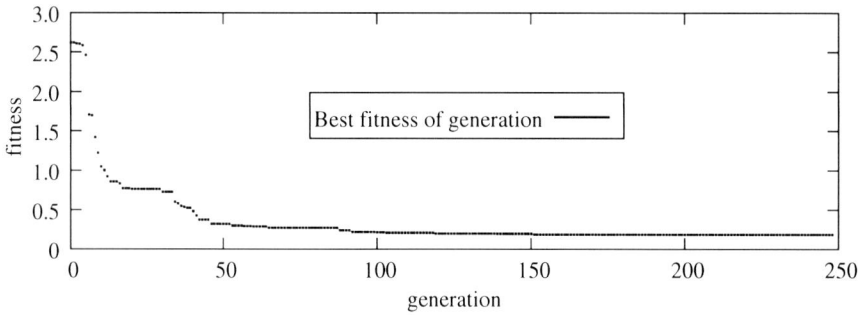

Fig. 6.8 Difference to target pattern (so a low value is good) of the best individual over time in a typical run evolving French Flag organisms. Note that the first 15 generations or so are dominated by individuals simply somehow expressing all 3 colours - remember that was a trick: When pattern difference was the only fitness measure the EA gets often stuck, so the raw difference is multiplied by (4 - number of colours present) - giving a huge initial fitness advantage to individuals expressing all colours. Only afterwards the real pattern matching is advantageous. After generation 150 (best!) fitness does not increase further in this run.

Also note fig. 6.7, where morphogen production was disabled (only secretion of morphogen, not GRN dynamics) to check whether individuals evolved to rely on morphogen cues for conditional differentiation. Apparently the individual uses autonomous cues for a first symmetry break (as red and blue cells can be seen), but requires morphogen to fully differentiate.

6.3.2.1 Robustness, Scalability and Heterotropy

Interesting from a robustness point of view was development when perturbed with targeted cell deaths. As the model had a border at which growth stopped automatically, cells' of all evolved individuals were trying to grow and divide continuously – so unsurprisingly removed cells were replaced quickly. However the patterning would not always develop correctly or recover, especially when cell death occurred early in development. Some cells apparently act as "organisers" whose functionality could not be taken over by duplicates of other cells.

As pointed out earlier, a major motivation for the interest in developmental GRNs is scalability and modularity. These mechanisms allow for quick adaptation on short evolutionary timescales. This is what was found in most evolved "French Flag" organisms. Within a couple of generations[7] a good match of the flag pattern twice the original size (120×40 instead of 60×40) could be achieved, see fig. 6.9 for an example. While the relative pattern was held constant (three stripes of equal width), the new setting required more than twice as many cells to fill the space.

In a follow-up experiment it was found that also a change in stripe ratios (from 33.3% for each of the three colours to 50%, 25%, 25%) of the pattern while doubling

---

[7] Note that, as before, the number of genes was kept constant during the additional evolution.

Fig. 6.9 French Flag after further evolving for a 120×40 pattern. Top: Target pattern. Middle/Bottom: Example individual after 15 additional generations, at $t = 200$, with 318 cells. (Expressed colour/Morphogen level)

size was possible, although this took more generations and resulted in a less perfect match on average, see fig. 6.10 for an example.

6.3.3 Discussion

The proposed model combines for the first time spatially extended cells with GRN control over individual cell parameters like size, shape, adhesion, morphogen secretion and orientation without predefined cell types. Evolved multicellular organisms were able to autonomously set up an asymmetric morphogen gradient and organise into a close match of the French flag pattern, so the model will certainly be of good use for further research. Due to the inherent randomness in the spatial process

Fig. 6.10 French Flag after further evolving for a 120×40 pattern with stripe fractions changed to 50%, 25%, 25% respectively. Example individual after 60 additional generations, at $t = 200$, with 308 cells. (Expressed colour/Morphogen level)

one can probably not expect a perfect pattern match on such a low level – nobody expects identical twins to be identical down to the cell level.

With only a couple of additional generations, the flag pattern and underlying gradient could be scaled to twice the size the individual was originally evolved for, using more than twice the number of cells. Similarly, the relative sizes of the modules could be changed, as experiments showed where the three equally sized stripes were further evolved to yield 50%, 25%, 25% stripe ratios (while also doubling target size). However this took more generations than the scaling alone, and boundaries between stripes where not as clear.

The fact that the initial cell in the first time step had an absolute orientation of zero degrees is clearly biologically unrealistic. The reason for doing it was simply that this made assessing fitness easier. This is not a major drawback however, as when the initial cell was started with an orientation of 90 degrees and the grid and target pattern were rotated by 90 degrees as well, the similarity achieved was just as good on average. Another unrealistic restriction is the fixed volume necessary to initiate mitosis; cell division should be triggered dynamically by a protein level.

On the timescales simulated here, cell adhesion could not and did not play a very important role. This was tested by turning off adhesion in evolved flag organisms – which generally resulted only in a small fitness drop. In more complex settings one should make sure to increase time scales so that evolution can take full advantage of adhesion, a factor developmental biologists think very important in morphogenesis.

Furthermore a search for evolved "cell types" was not very fruitful – trying to cluster protein expression patterns did not lead to strong clusters. Probably the evolutionary task was too simple: Timing played a big role and the target was only pattern formation without further functionality of the cells. The latter is a crucial issue, as the evolutionary target directly rewarded "modularisation". In nature however, this is a byproduct and only selected for if it conveys functional (ultimately reproductive) advantages.

In the future the evolution of morphogenesis can be extended to more complex, three dimensional forms instead of 2D patterns (as the name CompuCell3D suggests this is relatively straightforward with the software used).

Chapter 7
Conclusions

7.1 Summary

This research set out to investigate the evolvability of computational genetic regulatory networks (GRNs). As a basis, chapter 2 reviewed evolution in nature and genetic algorithms as computational abstraction of the natural process. The concept of evolvability was discussed in the light of natural and artificial evolution, with a particular focus on modularity and "duplication and divergence". The introductory part was completed by a description of biological GRNs and a literature review of GRN models. Against this background, the xBioSys GRN model and evolutionary setup, used throughout this work, was introduced. It allows any number of genetic binding sites per gene, realised through variable genome length. These inputs can be combined in two regulatory levels, trying to capture synergistic effects of transcription factors (TFs). Additionally, xBioSys features "smooth matching" of TFs to genetic binding sites with variable affinities between the two, dynamically controlled by specificity factors.

The performance of the xBioSys system was evaluated empirically in various evolutionary settings. Periodic activity, as in "biological clocks", and differentiation as the basis for division of labour are identified as crucial cell behaviours and thus useful benchmarks for GRN models. Biological clocks have probably been present in earliest single celled life and are characteristic for GRNs as adaptive control systems. In section 4.3 GRNs were evolved to respond to periodic environmental stimuli with varying periodic behaviours. Environmental stimulation came with and without (Gaussian) noise and "black out periods". Robustness to certain types of perturbation appears to account for some, but not all, dynamical properties of the evolved networks. Unselected abilities, also observed for biological clocks, include the capacity to adapt to change in wavelength of environmental stimulus, and to clock resetting.

Cellular differentiation, the switching between different control dynamics, was the evolutionary target in a next step. GRNs were evolved to produce different kinds of periodic dynamics in response to an inducing signal. Evolvability was shown to depend on evolutionary conditions and model features: Changing the environment over evolutionary time, allowing a lineage to adapt gradually to demands for differentiation was found to yield better results on average than static environments with pressure for differentiation from the beginning. Apart from better success there was also less variability in performance, the latter indicating an increase in evolutionary robustness. Furthermore xBioSys with smooth protein matching (plus specificity factors) was compared to a more classical variant without this mechanism. On evolutionary timescales, in the latter version mutations tend to completely change regulatory influences while with smooth matching the strength of influence is only slightly shifted. For evolutionary search spaces of increasing sizes evolved performance dropped much more strongly in the classical network model as compared to the smooth binding model. This effect was even greater in the case of using smooth binding together with specificity factors.

In chapter 5 the topological structure of networks was analysed. Hoping that small patterns with particular functionality would reliably evolve, two groups of GRNs were compared: Networks capable of exhibiting two different behaviours in contrast to a group with a single target behaviour. However, in both groups motif distribution differences within the groups were found to be larger than differences between them. This indicates that evolutionary niches (target functions) do not necessarily mold network structure uniquely. These results also showed that variability operators can have a stronger influence on network topologies than selection pressures, especially when many topologies can create similar dynamics. Moreover, analysis of motif functional relevance by lesioning did not suggest that motifs were of greater importance to the functioning of the network than arbitrary sub-graph patterns. Only when drastically restricting network size, so that one motif corresponds to a whole functionally evolved network, was preference for particular connection patterns found. This suggests that in non-restricted, bigger networks, entanglement with the rest of the network hinders topological sub-graph analysis.

Multicellular development became the focus in chapter 6, starting with a discussion of the physical dimension of morphogenesis and its importance. Particular weight was put on the self-organised formation of spatio-temporal patterns of differentiated cells. A review of morphogenesis inspired models followed in section 6.2. The Cellular Potts Model (CPM) was described and identified as providing an appropriate trade-off between physical realism and computational tractability. CPM contains a notion of energy constraints and allows arbitrary, heterogeneous cell shapes. Energy constraints were used to mimic cell adhesion, shape and growth. GRN-controlled protein levels were mapped to cell parameters such as energy constraints and protein diffusion. In experiments with the model, GRNs were successfully evolved to display "French flag" patterns. Self-organised differentiation (as opposed to externally induced differentiation) for spatial patterning and scalability of the modular patterns was found.

7.2 Review of Contributions

Revisiting the initial research questions and objectives of this work the following conclusions are drawn:

A computational genetic regulatory network (GRN) model that uses principles overlooked or considered irrelevant before, such as synergistic interaction between regulatory factors, has been introduced. Networks that respond like biological clocks to periodic environmental stimuli could be evolved from random initial network configurations. Analysis revealed internalisation of stimuli and phase resetting behaviour (which was not selected for), similar to the "incessant responsiveness" observed in natural organisms [West-Eberhard(2003)]. The xBioSys model has also been demonstrated to have characteristics like adaptability, robust control, potential for differentiation in common with its biological counterparts.

This work investigated and identified mechanisms that systematically influence evolvability. Variable fitness functions, gradually promoting differentiation over evolutionary time, led to better outcomes. Additionally lower variance of outcomes was found, meaning increased independence from the random start population and thus higher evolutionary robustness. Reviewing natural GRNs it was found that they should not be viewed as static, rather GRNs dynamically change regulatory affinities and their topology (e.g. regulated alternative splicing can change a gene's product) – not only on evolutionary time scales but during an organism's lifetime. In the xBioSys model variable affinities, controlled by gene products termed specificity factors, were shown to improve evolvability in large evolutionary search spaces, albeit at a computational cost.

Although suggested by [Alon(2006)] and others, attempts of breaking down network complexity by identifying recurring sub-graph pattern modules with particular functionality failed. However, this failure implies that generally partitioning networks is a poor basis for drawing conclusions about the dynamics of the system: Different configurations may generate similar dynamics and entanglement with the rest of the network may crucially influence function. While isolated motifs (i.e. with no or few inputs and outputs) may exhibit particular functionality, generally topology was found not to be a good predictor for function.

The importance of embodiment and physical constraints for biological morphogenesis was shown and an appropriate spatial model, with a good balance of computational tractability and abstraction, introduced. Spatially extended cells with GRN control over individual cell parameters like size, shape, adhesion, morphogen secretion and orientation without predefined cell types or induction were placed in such environments. Evolved individuals could mediate behaviour and differentiation among the constituent cells. Such multicellular organisms were able to autonomously set up an asymmetric morphogen gradient and self-assemble into a close match of the French flag pattern. Finally it was demonstrated that such modular patterns can be scaled up relatively easily (while keeping the number of genes constant).

This work has contributed to deepening our understanding of the evolvability of genetic regulatory networks and computational models thereof. While the work does not suggest that they are ideal for all applications, evidence has been provided that *computational models of genetic regulatory networks exhibit a high degree of evolvability*.

7.3   Reflection and Future Directions

The proposed GRN model xBioSys and its embodiment in the spatial CPM system have been demonstrated useful and thus lend themselves to further investigation. This last section seeks to reflect on the work and point out possible future directions for understanding evolvability in the computational modelling of GRNs and development.

Modelling synergistic regulation and dynamic affinities proved to be a good step in the direction of more evolvable GRN logics. Other mechanisms have not been modelled although evolutionary biologists suspect they have a strong influence on the evolvability of natural GRNs.

Alternative splicing for example, i.e. gene products whose production process can be modified to some degree dynamically by transcription factors, thereby dynamically changing GRN topology (see section 3.1 for details), might be an interesting extension for future versions. Many biologists, e.g. [Sammeth et al(2008)Sammeth, Foissac, and Guigo], assume that alternative splicing has allowed Eukaryotes to adapt faster. With an appropriate model some light could be shed on this claim.

The synchronous update regime in xBioSys and many other models might have an impact on system behaviour, which should be quantified. Asynchronous, non-deterministic updates may be more realistic and robust as research by others indicates (cf. section 3.2). Still, interactions in such systems (as well as in natural GRNs of course) are not totally random, as different processes act on different timescales. This makes it very likely that some events will occur before others, resulting in a certain predictability even in the face of noisy fluctuations.

On evolutionary timescales we have seen average population performance staying pretty much the same after strong initial improvements. This is probably at least partly due to a too destructive crossover, as genomes get longer and longer, having lots of junk and dysfunctional binding sites and genes. In the future this might be changed by assigning higher probabilities to particular points for crossover and offset instead of random ones, leading to a more meaningful crossover. There seems to be consent that unequal crossover in biological GRNs is not totally random but that there are preferred crossover places. It is not known where exactly these preferred points are, but if they would for example be between binding sites instead of inside them this might make crossover less destructive. Additionally there are no costs on genome length, so there is no disadvantage to long genomes. The reasons are probably similar to the finding of Koza from Genetic Programming that programs tend to become redundant as this provides some protection from often harmful crossover effects (buffer-like).

Early experiments with a gene duplication operator in xBioSys have produced mediocre outcomes[1]. Duplications would probably have more of a positive impact if neutral duplications were possible – in the present xBioSys model this is only the case if regulation is in the saturation area of the sigmoid, so that dividing input protein levels by two still results in maximum output. Additionally it is worth considering that duplication can take place either completely, with all of a gene's binding sites, or copying the binding sites with some probability only. The latter might biologically be more realistic as sites with regulatory effects can be very far away from the coding region of the gene and thus are not necessarily included in a duplication event. Another factor is probably meaningful crossover between genomes with different gene numbers. In nature it is known that similar chromosomes do align, but it is not known how exactly similarity is determined.

Attempts to investigate structural modularity of individual GRNs and the influence of structure on function using motif analysis revealed the non-existence of such a relationship based on simple motifs. Still, a better understanding of dynamical and structural composition of GRN would make it easier to construct GRNs with particular functionality. Hierarchical algebraic decomposition [Egri-Nagy and Nehaniv(2008)] might shed light on meaningful genetic units ("building block hypothesis") and on how the organisation of GRN clocks is affected by their evolutionary history. Another promising candidate could be "activity motifs" by Daphne Koller's group, see e.g. [Chechik et al(2008)Chechik, Oh, Rando, Weissman, Regev, and Koller] – who try to link structural and dynamical system properties. Generally it is desirable to develop better analytical tools for dynamical systems both on the level of GRNs as well as cellular networks.

The "French flag" organisms achieve emergent differentiation and modularisation, but how is this reflected in the corresponding GRN dynamics and topology? First insights may be gained by gene knock-out experiments, as frequently used by molecular biologists today. A systematic analysis of their self-organisation, possibly by applying information theoretic measures to the signalling between the cells, would be interesting. Generally much more work is needed to understand the dynamics of multicellular "networks of networks".

First however different evolutionary settings should be examined. Self-repair properties after perturbations to development would be expected as for example [Miller(2004)] observed in his work. Particularly, perturbations should be experienced during evolution, as this would encourage the use of robust developmental mechanisms in the flag-organisms. A modular organisation has been found in the experiments reported here. Although these patterns could be quickly evolved to scale and change their relative sizes, it still has to be shown that further evolution can make use of duplication and divergence of the modules. Finding developmental modules that can be duplicated and later diverge to allow for modulated repetition (figure 2.2 on page 10) would be very powerful. Early (non-systematic) experiments have shown that evolved indivduals made use of boundary effect, which is likely due

---

[1] These experiments have not been published, but results are available online - at http://panmental.de/CECdynAff/#moreExperiments

to the fixed grid size experienced during evolution. This caused problems as soon as the grid was changed, for example to accomodate additional stripe segments. Accordingly such behaviour should be avoided in future experiments by either growing indivduals on a fixed size grid with a margin around the target pattern or by using a grid that expands dynamically with the cells.

In the future more natural fitness measures should be used, evolving for function with multicellular shape as a by-product rather than selecting for shape directly. This might be achieved by extending the model with a notion of "food", an energy source that has to be consumed by individuals in order to grow and divide – as there is a chemotaxis module in the CompuCell3D package, introduction of movement is straightforward. If modularity is still found in a setting where it is not selected for this would be stronger evidence that it is an evolutionary advantageous organisation principle. Of course active cell movement or motility would also merit to be included in its own right, as it is at least as important for morphogenesis as cells' control of shape.

Furthermore, biological cells are even more flexible and irregular in the forms they can take; for example it is known that contraction, leading to a wedge like shape, is important for gastrulation [Wolpert et al(2007)Wolpert, Jessell, Lawrence, Meyerowitz, Robertson, and Smith]. It would be interesting to see if evolvability is increased further when the GRN gets more control over cell shape.

An engineering application where a variant of xBioSys could be useful in the near future is the generation of neural systems. The growth of neural systems is a hard, not straightforward optimisation task – but while it is not clear what the optimal layout is, the brain is certainly highly modular. Additionally the importance of changes to gene activity for long term learning, during an organisms lifetime, are becoming increasingly clear. Mechanisms under GRN control include synaptic plasticity and changes in general excitability, but also morphological processes like growth of new synapses and neurogenesis. For more details see e.g. [Kandel(2001), West et al(2002)West, Griffith, and Greenberg].

Most applications of evolutionary algorithms to the design of neural systems so far have used classical neural networks and focussed on optimising parameters determining structure and synaptic weight. Some developmental GRN models have already been used to evolve neural networks (see discussion in 3.2). However, these models had a very abstract growth process, neglecting dynamical self-organisation and environmental interaction, and GRNs were usually discarded after the neural system was created.

Generally, it would be a good idea for researchers in this area to cooperate more. GRN models may differ in the evolvability of expressive regulatory dynamics. As the source code is rarely made available and descriptions are often not sufficient for a full reimplementation it would at least be desirable to have a standard test set to compare models, especially with regard to evolvability. This might also help in finding the best possible trade-off between model detail and size of the search space.

Appendix A
Pseudo Code

In this appendix pseudo code for the GRN update algorithm of the xBioSys model, described in section 3.3, is presented.

Full source code is available at:

http://panmental.de/ALifeXIflag/ (Development and Morphogenesis)
http://panmental.de/motifs/ (Topological Network Analysis)
http://panmental.de/GRNclocks/ and http://panmental.de/CECdynAff/ (Dynamical Clocks and Differentiation)

A.1 GRN Processing

```
//per time step
function process(){ //perfect matching version
//1. bind free proteins to matching cis sites
foreach GP : freeGPs{
  sharePerSite=GP.amount*bindingProportion/
                        sitesReceptiveFor(GP.type).length;
  foreach site : sitesReceptiveFor(GP.type)
    site.bind(sharePerSite);
  GP.addAmount(-(GP.amount*bindingProportion));
  }
//2. call output function of genes and add output to free GP
foreach GP : freeGPs
    foreach gene : genes
    GP.addAmount(gene.getOutputFor(GP));
//3. saturation (upper bound) check
foreach GP : freeGPs
    if(GP.amount>saturationValue)
        GP.amount=saturationValue;
//4. decay free GP according to protein specific decay rate
foreach GP : freeGPs
  GP.amount=GP.amount*GP.proteinDecay;
```

```
//[5. output and input using scaling factor r]
}

function getBellCurveVal(double x, double sigma){
    return (1.0/sigma*root(2*PI))*exp(-((x*x)/(2*sigma*sigma))));
}

//...
function process(){ //smooth matching version
//1. bind free proteins to matching cis sites
foreach GP : freeGPs {
    shares=Array(n); //n is the number of bits for a GP
    shareSum=0.0;
    for(x=0;x<shares.length;x++){
    shares[x]=getBellCurveVal(x,0.5+GP.specificityFactor);
    shareSum+=shares[x];
        }
    for(int x=0;x<shares.length;x++) //normalize
    shares[x]/=shareSum;
    for(prot=0;prot<freeGPs.length;prot++){
    dis=getHammingDistance(toBinary(GP.type),toBinary(prot));
    share=shares[dis];
        switch(dis){
            case 0: num=1.0; break; //only one perfect match, i.e. type==prot
            //... how many proteins have hamming distance dis, up to n bits
        case n: num=1.0; break; //one differs in all three bits
        }
      amountForAllWithThisDis=(GP.amount*bindingProportion)*share;
    foreach site : sitesReceptiveFor(GP.type)
        site.bind(amountForAllWithThisDis/num/
                            sitesReceptiveFor(GP.type).length);
    }
    GP.amount-=(GP.amount*bindingProportion);
}
//2. (see above)...
```

A.2   Gene Processing

```
function tanhScaled(x, range, floor, rshift, slope){
//floorScale is added at the end to shift the whole 's' curve up.
floorScale = floor + (range/2);
//By reducing value of x, the whole 'S' is shifted right.
x = x - rshift;
return ( (((exp(x/slope) - exp(-x/slope)) / 2) /
         ((exp(x/slope) + exp(-x/slope)) / 2)
         )
         * (range/2)
       ) + floorScale;
}

public double getOutputFor(GP){
if(GP.type!=this.type)
    return 0;
activation=0;
foreach cisModule : cisModules {
    if(cisModule.bindingSites.length==0)
        continue;
    temp=999999.0;
    foreach bindingSite : cisModule.bindingSites {
        if(bindingSite<temp)
        temp=bindingSite;
```

```
        bindingSite*=GP.proteinDecay; //decay bound GP
        }
  if(cisModule.regulationMode)      //inhibitory
     activation-=temp;
  else                             //ativatory
     activation+=temp;
  }
if(expressionType) //default 'off' gene
  shift=15.0;
else                              //default 'on' gene
  shift=-5.0;
return tanhScaled(activation,150,0.0,shift,5);
}
```

Appendix B
Further Reading

In parts, the research presented in this book is based on earlier publications, including my PhD thesis, a book chapter, two journal, and six publications in conference proceedings.

Knabe, J. F., Wegner, K., Nehaniv, C. L. and Schilstra, M. J. Genetic Algorithms and Their Application to In Silico Evolution of Genetic Regulatory Networks. In Fenyo, D (ed) *Computational Biology*, Vol. 673, pages 297-321, Humana Press, 2010.

Knabe, J. F. *Evolvability of Computational Genetic Regulatory Networks*. PhD thesis, Science and Technology Research Institute, University of Hertfordshire, Hatfield, Hertfordshire, UK, 2009.

Knabe, J. F., Nehaniv, C. L. and Schilstra, M. J. Do Motifs Reflect Evolved Function? – No Convergent Evolution of Genetic Regulatory Network Subgraph Topologies. In *BioSystems*, 94 (1-2): 68-74, 2008.

Knabe, J. F., Schilstra, M. J. and Nehaniv, C. L. Evolution and Morphogenesis of Differentiated Multicellular Organisms: Autonomously Generated Diffusion Gradients for Positional Information. In *Artificial Life XI: Proceedings of the Eleventh International Conference on the Simulation and Synthesis of Living Systems*, pages 321-328, MIT Press, 2008.

Knabe, J. F., Nehaniv, C. L. and Schilstra, M. J. Regulation of Gene Regulation - Smooth Binding with Dynamic Affinity affects Evolvability. In *IEEE Congress on Evolutionary Computation (CEC 2008). Proc. WCCI 2008*, pages 890-896, IEEE Press, 2008.

Knabe, J. F., Nehaniv, C. L. and Schilstra, M. J. Genetic Regulatory Network models of Biological Clocks: Evolutionary history matters. In *Artificial Life*, 14 (1): 135-148, 2008.

Knabe, J. F., Nehaniv, C. L. and Schilstra, M. J. Unruly Motifs – No Convergent Evolution of Network Topologies. In *International Workshop on Information Processing in Cells and Tissues (IPCAT)*, pages 182-191, Librix, 2007.

Knabe, J. F., Nehaniv, C. L. and Schilstra, M. J. The Essential Motif that wasn't there: Topological and Lesioning Analysis of Evolved Genetic Regulatory

Networks. In *IEEE Symposium on Artificial Life (CI-ALife'07)*, pages 69-76, Omnipress, 2007.

Knabe, J. F., Nehaniv, C. L. and Schilstra, M. J. Evolutionary Robustness of Differentiation in Genetic Regulatory Networks. In *Proceedings of the 7th German Workshop on Artificial Life 2006 (GWAL-7)*, pages 75-84, Akademische Verlagsgesellschaft Aka, Berlin, 2006.

Knabe, J. F., Nehaniv, C. L., Schilstra, M. J. and Quick, T. Evolving Biological Clocks using Genetic Regulatory Networks. In *Artificial Life X: Proceedings of the Tenth International Conference on the Simulation and Synthesis of Living Systems*, pages 15-21, MIT Press/Bradford Books, 2006.

References

[Albert and Othmer(2003)] Albert, R., Othmer, H.G.: The topology of the regulatory interactions predics the expression pattern of the segment polarity genes in drosophila
melanogaster. Journal of Theoretical Biology 223 (2003)

[Alberts et al(2002)Alberts, Johnson, Lewis, Raff, Roberts, and Walter] Alberts, B., Johnson, A., Lewis, J., Raff, M., Roberts, K., Walter, P.: Molecular Biology of the Cell,
4th edn. Garland Science, New York and London (2002)

[Alon(2006)] Alon, U.: An Introduction to Systems Biology – Design Principles of Biologial
Circuits. Chapman & Hall/CRC (2006)

[Altenberg(1994)] Altenberg, L.: The evolution of evolvability in genetic programming.
In: Kinnear, K.E. (ed.) Advances in Genetic Programming, pp. 47–74. MIT Press,
Cambridge (1994)

[Altenberg(1995)] Altenberg, L.: Genome growth and the evolution of the genotype-
phenotype map, pp. 205–259. Springer (1995)

[Arthur(2000)] Arthur, W.: The Origin of Animal Body Plans, paperback edn. Cambridge
University Press, Cambridge (2000)

[Artzy-Randrup et al(2004)Artzy-Randrup, Fleishman, Ben-Tal, and Stone] Artzy-
Randrup, Y., Fleishman, S.J., Ben-Tal, N., Stone, L.: Comment on "Network Motifs:
Simple Building Blocks of Complex Networks" and "Superfamilies of Evolved and
Designed Networks". Science 305(5687), 1107c (2004), doi:10.1126/science.1099334

[Back et al(1999)Back, Fogel, and Michalewicz] Back, T., Fogel, D.B., Michalewicz, Z.
(eds.): Evolutionary Computation 1: Basic Algorithms and Operators. IOP Publishing Ltd., Bristol (1999)

[Baldwin(1896)] Baldwin, J.M.: A new factor in evolution. American Naturalist 30, 441–
451, 536–553 (1896)

[Banzhaf(2003)] Banzhaf, W.: On the Dynamics of an Artificial Regulatory Network. In:
Banzhaf, W., Ziegler, J., Christaller, T., Dittrich, P., Kim, J.T. (eds.) ECAL 2003.
LNCS (LNAI), vol. 2801, pp. 217–227. Springer, Heidelberg (2003)

[Beer(1994)] Beer, S.: Brain of the Firm. Classic Beer Series. Wiley (1994)

[Bentley(2004)] Bentley, P.J.: Adaptive fractal gene regulatory networks for robot control.
In: Miller, J. (ed.) Workshop on Regeneration and Learning in Developmental Systems, Genetic and Evolutionary Computation Conference, GECCO 2004 (2004)

[Bongard(2002)] Bongard, J.: Evolving modular genetic regulatory networks. In: Proc. of
the Congress on Evolutionary Computation 2002, pp. 1872–1877. IEEE Computer
Society, Washington, DC (2002)

[Bongard and Pfeifer(2001)] Bongard, J.C., Pfeifer, R.: Repeated structure and dissociation
of genotypic and phenotypic complexity in artificial ontogeny. In: Spector, L., et al.
(eds.) Proc. of the Genetic and Evolutionary Computation Conference (GECCO 2001),
pp. 829–836. Morgan Kaufmann, San Francisco (2001)

[Bonner(2001)] Bonner, J.T.: First Signals: The Evolution of Multicellular Development. Princeton University Press, Princeton (2001)

[Bornholdt(2005)] Bornholdt, S.: Systems biology: Less is more in modeling large genetic networks. Science 310(5747), 449–451 (2005),
doi: http://dx.doi.org/10.1126/science.1119959

[Buck and Nehaniv(2008)] Buck, M., Nehaniv, C.L.: Communication and complexity in a grn-based multicellular system for graph colouring. Biosystems 94(1-2), 28–33 (2008); seventh International Workshop on Information Processing in Cells and Tissues (IPCAT 2007) (2007)

[Buss(1987)] Buss, L.W.: The Evolution of Individuality. Princeton University Press, Princeton (1987)

[Calabretta et al(2000)Calabretta, Nolfi, Parisi, and Wagner] Calabretta, R., Nolfi, S., Parisi, D., Wagner, G.P.: Duplication of modules facilitates the evolution of functional specialization. Artificial Life 6(1), 69–84 (2000)

[Carroll(2008)] Carroll, S.B.: Evo-devo and an expanding evolutionary synthesis: A genetic theory of morphological evolution. Cell 134(1), 25–36 (2008),
doi: http://dx.doi.org/10.1016/j.cell.2008.06.030

[Carroll et al(2001)Carroll, Grenier, and Weatherbee] Carroll, S.B., Grenier, J.K., Weatherbee, S.D.: From DNA to Diversity: Molecular Genetics and the Evolution of Animal Design. Blackwell Science (2001)

[Chaves et al(2005)Chaves, Albert, and Sontag] Chaves, M., Albert, R., Sontag, E.D.: Robustness and fragility of boolean models for genetic regulatory networks. Journal of Theoretical Biology 235, 431–449 (2005)

[Chechik et al(2008)Chechik, Oh, Rando, Weissman, Regev, and Koller] Chechik, G., Oh, E., Rando, O., Weissman, J., Regev, A., Koller, D.: Activity motifs reveal principles of timing in transcriptional control of the yeast metabolic network. Nature Biotechnology, 1251–1259 (2008)

[Cickovski et al(2007)] Cickovski, T., Aras, K., Swat, M., Merks, R.M.H., Glimm, T., Hentschel, H.G.E., Alber, M.S., Glazier, J.A., Newman, S.A., Izaguirre, J.A.: From genes to organisms via the cell: A problem-solving environment for multicellular development. Computing in Science and Eng. 9(4), 50–60 (2007)

[Conant and Wagner(2003)] Conant, G.C., Wagner, A.: Convergent evolution of gene circuits. Nature Genetics 34(3), 264–266 (2003)

[Conrad(1990)] Conrad, M.: The geometry of evolution. Biosystems 24(1), 61–81 (1990)

[Cordero and Hogeweg(2006)] Cordero, O., Hogeweg, P.: Feed Forward Loop Circuits as a Side Effect of Genome Evolution. Mol. Biol. Evol. 23, 1931–1936 (2006)

[Darwin(1859)] Darwin, C.: On the Origin of Species. John Murray, London (1859)

[Davidson(2001)] Davidson, E.H.: Genomic Regulatory Systems: Development and Evolution. Academic Press, Burlington (2001)

[Davidson(2006)] Davidson, E.H.: The Regulatory Genome: Gene Regulatory Networks in Development and Evolution. Academic Press (2006)

[de Jong(1975)] de Jong, K.A.: An analysis of the behaviour of a class of genetic systems. PhD thesis. University of Michigan (1975)

[Dellaert and Beer(1996)] Dellaert, F., Beer, A.D.: A developmental model for the evolution of complete autonomous agents. In: Proceedings of the Fourth International Conference on Simulation of Adaptive Behavior, pp. 393–401. MIT Press (1996)

[Devert et al(2007)Devert, Bredeche, and Schoenauer] Devert, A., Bredeche, N., Schoenauer, M.: Robust multi-cellular developmental design. In: GECCO 2007: Proceedings of the 9th Annual Conference on Genetic and Evolutionary Computation, pp. 982–989. ACM, New York (2007)

[Dobzhansky(1973)] Dobzhansky, T.: Nothing in biology makes sense except in the light of evolution. American Biology Teacher 35, 125–129 (1973)

[Drennan and Beer(2006)] Drennan, B., Beer, R.D.: Evolution of repressilators using a biologically-motivated model of gene expression. In: Rocha, L.M., Yaeger, L.S., Bedau, M.A., Floreano, D., Goldstone, R.L., Vespignani, A. (eds.) Artificial Life X: Proceedings of the Tenth International Conference on the Simulation and Synthesis of Living Systems. International Society for Artificial Life, pp. 22–27. The MIT Press, Bradford Books (2006)

[Eggenberger(2003)] Eggenberger, P.: Genome-physics interaction as a new concept to reduce the number of genetic parameters in artificial evolution. In: Proceedings of the IEEE 2003 Congress on Evolutionary Computation, Canberra, Australia, pp. 191–198 (2003)

[Eggenberger-Hotz(1997)] Eggenberger-Hotz, P.: Evolving morphologies of simulated 3d organisms based on differential gene expression. In: Husbands, P., Harvey, I. (eds.) Proceedings of the 4th European Conference on Artificial Life (ECAL 1997). MIT Press, Cambridge (1997)

[Egri-Nagy and Nehaniv(2008)] Egri-Nagy, A., Nehaniv, C.L.: Hierarchical coordinate systems for understanding complexity and its evolution, with applications to genetic regulatory networks. Artificial Life (Special Issue on Evolution of Complexity) 14(3), 299–312 (2008)

[Federici(2004)] Federici, D.: Increasing evolvability for developmental programs. In: Miller, J. (ed.) Proceedings of the Workshop on Regeneration and Learning in Developmental Systems, WORLDS 2004 (2004)

[Federici and Ziemke(2006)] Federici, D., Ziemke, T.: Why are evolved developing organisms also fault-tolerant? In: Nolfi, S., Baldassarre, G., Calabretta, R., Hallam, J.C.T., Marocco, D., Meyer, J.-A., Miglino, O., Parisi, D. (eds.) SAB 2006. LNCS (LNAI), vol. 4095, pp. 449–460. Springer, Heidelberg (2006)

[Fleischer and Barr(1993)] Fleischer, K., Barr, A.H.: A simulation testbed for the study of multicellular development: The multiple mechanisms of morphogenesis. In: Langton, C.G. (ed.) Artificial Life III, pp. 389–416. Addison-Wesley (1993)

[Fogel(1998)] Fogel, D.B.: Evolutionary Computation: The Fossil Record. Wiley-IEEE Press (1998)

[Fogel et al(1966)Fogel, Owens, and Walsh] Fogel, L.J., Owens, A.J., Walsh, M.J.: Artificial Intelligence through Simulated Evolution. John Wiley, New York (1966)

[Folkman and Moscona(1978)] Folkman, J., Moscona, A.: Role of cell shape in growth control. Nature 273(5661), 345–349 (1978)

[Forgacs and Newman(2005)] Forgacs, G., Newman, S.: Biological Physics of the Developing Embryo. Cambridge University Press (2005)

[Furusawa and Kaneko(1998)] Furusawa, C., Kaneko, K.: Emergence of multicellular organisms with dynamic differentiation and spatial pattern. Artificial Life 4(1), 79–93 (1998)

[Geard and Wiles(2005)] Geard, N., Wiles, J.: A gene network model for developing cell lineages. Artificial Life 11(3), 249–268 (2005), doi:10.1162/1064546054407202

[Gerhart and Kirschner(1997)] Gerhart, J.C., Kirschner, M.: Cells, Embryos and Evolution. Blackwell Publishing (1997)

[Gershenson(2002)] Gershenson, C.: Classification of random boolean networks. In: Standish, R.K., Bedau, M.A., Abbass, H.A. (eds.) Artificial Life VIII: Proceedings of the Eight International Conference on Artificial Life, pp. 1–8. MIT Press (2002)

[Glazier et al(2005)Glazier, Heiland, Swat, and Zaitlen] Glazier, J., Heiland, R., Swat, M., Zaitlen, B.: Compucell3d software package and manual (2005), http://compucell3d.org/

[Glazier and Graner(1993)] Glazier, J.A., Graner, F.: Simulation of the differential adhesion driven rearrangement of biological cells. Phys. Rev. E. 47(3), 2128–2154 (1993), doi:10.1103/PhysRevE.47.2128

[Goldberg and Richardson(1987)] Goldberg, D.E., Richardson, J.: Genetic algorithms with sharing for multi-modal function optimization. In: Proceedings of the 2nd International Conference on Genetic Algorithms, pp. 41–49 (1987)

[Gould(1977)] Gould, S.J.: Ontogeny and Phylogeny. Belknap Press / Harvard University Press, Cambridge (1977)

[Gould and Lewontin(1979)] Gould, S.J., Lewontin, R.C.: The spandrels of San Marco and the Panglossian paradigm: a critique of the adaptationist programme. Proc. R. Soc. Lond. B. Biol. Sci. 205(1161), 581–598 (1979)

[Gregory(2004)] Gregory, R.T.: The Evolution of the Genome. Academic Press, Burlington (2004)

[Halder et al(1995)Halder, Callaerts, and Gehring] Halder, G., Callaerts, P., Gehring, W.J.: Induction of ectopic eyes by targeted expression of the eyeless gene in *Drosophila*. Science 267(5205), 1788–1792 (1995)

[Harvey(1992)] Harvey, I.: Species adaptation genetic algorithms: A basis for a continuing saga. In: Proceedings of the First European Conference on Artificial Life, pp. 346–354. MIT Press/Bradford Books (1992)

[Ho(2007)] Ho, J.W.K.: Modeling the Evolution of Gene Regulatory Networks. Presentation at The Eighth International Conference on Systems Biology 2007, Long Beach, CA (2007)

[Hogeweg(2000)] Hogeweg, P.: Shapes in the shadow: Evolutionary dynamics of morphogenesis. Artificial Life 6(1), 85–101 (2000)

[Holland(1975)] Holland, J.H.: Adaptation in Natural and Artificial Systems: An Introductory Analysis with Applications to Biology. Control, and Artificial Intelligence. University of Michigan Press, Ann Arbor (1975)

[Huang and Ingber(2000)] Huang, S., Ingber, D.E.: Shape-dependent control of cell growth, differentiation, and apoptosis: Switching between attractors in cell regulatory networks. Experimental Cell Research 261(1), 91–103 (2000), doi:10.1006/excr.2000.5044

[Huxley(1942)] Huxley, J.: Evolution: The Modern Synthesis. George Allen and Unwin, London (1942)

[Huynen et al(1996)Huynen, Stadler, and Fontana] Huynen, M.A., Stadler, P.F., Fontana, W.: Smoothness within ruggedness: the role of neutrality in adaptation. Proc. Natl. Acad. Sci. U S A 93(1), 397–401 (1996)

[Iguchi et al(2005)Iguchi, Kinoshita, and Yamada] Iguchi, K., Kinoshita, S., Yamada, H.: Rugged fitness landscapes of kauffman models with a scale-free network. Physical Review E (Statistical, Nonlinear, and Soft Matter Physics) 72(6), 061901 (2005)

[Isalan et al(2008)] Isalan, M., Lemerle, C., Michalodimitrakis, K., Horn, C., Beltrao, P., Raineri, E., Garriga-Canut, M., Serrano, L.: Evolvability and hierarchy in rewired bacterial gene networks. Nature 452(7189), 840–845 (2008), doi:10.1038/nature06847

[Jacob and Monod(1961)] Jacob, F., Monod, J.: Genetic regulatory mechanisms in the synthesis of proteins. J. Mol. Biol. 3, 318–356 (1961)

[Jacob and Monod(1963)] Jacob, F., Monod, J.: Genetic repression, allosteric inhibition, and cellular differentiation. In: Locke, M. (ed.) Cytodifferentiation and Macromolecular Synthesis, pp. 30–64. Academic Press, London (1963)

[Jaeger and Reinitz(2006)] Jaeger, J., Reinitz, J.: On the dynamic nature of positional information. Bio. Essays 28(11), 1102–1111 (2006), doi:10.1002/bies.20494

[Jakobi(1995)] Jakobi, N.: Harnessing morphogenesis. In: International Conference on Information Processing in Cells and Tissues, pp. 29–41 (1995)

[Joachimczak and Wrobel(2008)] Joachimczak, M., Wrobel, B.: Evo-devo in silico: a model of a gene network regulating multicellular development in 3d space with artificial physics. In: Bullock, S., Noble, J., Watson, R., Bedau, M.A. (eds.) Artificial Life XI: Proceedings of the Eleventh International Conference on the Simulation and Synthesis of Living Systems, pp. 297–304. MIT Press, Cambridge (2008)

[de Jong(2002)] de Jong, H.: Modeling and simulation of genetic regulatory systems: A literature review. Journal of Computational Biology 9(1), 67–103 (2002)

[Kandel(2001)] Kandel, E.R.: The Molecular Biology of Memory Storage: A Dialogue Between Genes and Synapses. Science 294(5544), 1030–1038 (2001)

[Katok and Hasselblatt(1995)] Katok, A., Hasselblatt, B.: Introduction to the Modern Theory of Dynamical Systems. Cambridge University Press, New York (1995)

[Kauffman et al(2004)Kauffman, Peterson, Samuelsson, and Troein] Kauffman, S., Peterson, C., Samuelsson, B., Troein, C.: Genetic networks with canalyzing boolean rules are always stable. Proc. Natl. Acad. Sci. USA 101(49), 17,102–17,107 (2004)

[Kauffman(1969)] Kauffman, S.A.: Metabolic stability and epigenesis in randomly constructed genetic nets. Journal of Theoretical Biology 22, 437–467 (1969)

[Kauffman(1993)] Kauffman, S.A.: The Origins of Order: Self-Organization and Selection in Evolution. Oxford University Press (1993)

[Kauffman and Smith(1986)] Kauffman, S.A., Smith, R.G.: Adaptive automata based on Darwinian selection. Phys. D 2(1-3), 68–82 (1986)

[Kimura(1983)] Kimura, M.: The neutral theory of molecular evolution. Cambridge University Press, Cambridge (1983)

[Kirschner and Gerhart(1998)] Kirschner, M., Gerhart, J.: Evolvability. Proc. Natl. Acad. Sci. USA 95(15), 8420–8427 (1998)

[Kitano(1990)] Kitano, H.: Designing neural networks using genetic algorithms with graph generation system. Complex Systems Journal 4, 461–476 (1990)

[Knabe et al(2006)Knabe, Nehaniv, Schilstra, and Quick] Knabe, J.F., Nehaniv, C.L., Schilstra, M.J., Quick, T.: Evolving biological clocks using genetic regulatory networks. In: Rocha, L.M., Yaeger, L.S., Bedau, M.A., Floreano, D., Goldstone, R.L., Vespignani, A. (eds.) Artificial Life X: Proceedings of the Tenth International Conference on the Simulation and Synthesis of Living Systems, pp. 15–21. MIT Press/Bradford Books (2006)

[Knabe et al(2010)Knabe, Wegner, Nehaniv, and Schilstra] Knabe, J.F., Wegner, K., Nehaniv, C.L., Schilstra, M.J.: Genetic algorithms and their application to in silico evolution of genetic regulatory networks. In: Fenyo, D. (ed.) Computational Biology, Methods in Molecular Biology, vol. 673, ch. 19, pp. 297–321. Humana Press, Totowa (2010)

[Koza(1992)] Koza, J.R.: Genetic programming: on the programming of computers by means of natural selection. MIT Press, Cambridge (1992)

[Kumar(2004)] Kumar, S.: Investigating computational models of development for the construction of shape and form. PhD thesis. University College London, London (2004)

[Kumar and Bentley(2003)] Kumar, S., Bentley, P. (eds.): On Growth, Form and Computers. Elsevier Academic Press, London (2003)

[Kuo et al(2006)Kuo, Banzhaf, and Leier] Kuo, P., Banzhaf, W., Leier, A.: Network topology and the evolution of dynamics in an artificial regulatory network model created by whole genome duplication and divergence. BioSystems 85(3), 177–200 (2006)

[Lodish et al(2004)Lodish, Baltimore, Berk, Zipursky, Matsudaira, and Darnell] Lodish, H.F., Baltimore, D., Berk, A., Zipursky, S.L., Matsudaira, P., Darnell, J.E.: Molecular Cell Biology, 5th edn. W.H. Freeman, New York (2004)

[Mahfoud(1995)] Mahfoud, S.W.: Niching Methods for Genetic Algorithms. Tech. rep., Illinois Genetic Algorithms Laboratory (IlliGAL), Report No. 95001 (1995)

[Marée and Hogeweg(2001)] Marée, A.F.M., Hogeweg, P.: How amoeboids self-organize into a fruiting body: Multicellular coordination in *Dictyostelium discoideum*. Proc. Natl. Acad. Sci. USA 98(7), 3879–3883 (2001)

[Margulis and Sagan(1997)] Margulis, L., Sagan, D.: Microcosmos: four billion years of evolution from our microbial ancestors. University of California Press (1997)

[Maynard Smith(1998)] Maynard Smith, J.: Shaping Life - Genes, Embryos and Evolution. Weidenfeld and Nicolson, London (1998)

[Maynard Smith and Szathmáry(1998)] Maynard Smith, J., Szathmáry, E.: The Major Transitions in Evolution, reprint edn. W.H. Freeman, New York (1998)

[Mazurie et al(2005)Mazurie, Bottani, and Vergassola] Mazurie, A., Bottani, S., Vergassola, M.: An evolutionary and functional assessment of regulatory network motifs. Genome Biology 6(4), R35 (2005), doi:10.1186/gb-2005-6-4-r35

[McGinnis et al(1984)McGinnis, Levine, Hafen, Kuroiwa, and Gehring] McGinnis, W., Levine, M.S., Hafen, E., Kuroiwa, A., Gehring, W.J.: A conserved dna sequence in homoeotic genes of the drosophila antennapedia and bithorax complexes. Nature 308(5958), 428–433 (1984), doi: http://dx.doi.org/10.1038/308428a0

[Meinhardt(1978)] Meinhardt, H.: Space-dependent cell determination under the control of a morphogen gradient. Journal of Theoretical Biology 74(2), 307–321 (1978), doi:10.1016/0022-5193(78)90078-4

[Mendel(1866)] Mendel, J.G.: Versuche über Pflanzenhybriden, Verhandlungen des naturforschenden Vereines in Brünn 4 (1866)

[Merks and Glazier(2005)] Merks, R.M.H., Glazier, J.A.: A cell-centered approach to developmental biology. Physica A: Statistical Mechanics and its Applications 352(1), 113–130 (2005)

[Merks et al(2006)Merks, Brodsky, Goligorsky, Newman, and Glazier] Merks, R.M.H., Brodsky, S.V., Goligorsky, M.S., Newman, S.A., Glazier, J.A.: Cell elongation is key to in silico replication of in vitro vasculogenesis and subsequent remodeling. Dev. Biol. 289, 44–54 (2006)

[Miller(2003)] Miller, J.F.: Evolving Developmental Programs for Adaptation, Morphogenesis, and Self-Repair. In: Banzhaf, W., Ziegler, J., Christaller, T., Dittrich, P., Kim, J.T. (eds.) ECAL 2003. LNCS (LNAI), vol. 2801, pp. 256–265. Springer, Heidelberg (2003)

[Miller(2004)] Miller, J.F.: Evolving a Self-Repairing, Self-Regulating, French Flag Organism. In: Deb, K., Poli, R., Banzhaf, W., Beyer, H., Burke, E.K., Darwen, P.J., Dasgupta, D., Floreano, D., Foster, J.A., Harman, M., Holland, O., Lanzi, P.L., Spector, L., Tettamanzi, A., Thierens, D., Tyrrell, A.M. (eds.) GECCO 2004. LNCS, vol. 3102, pp. 129–139. Springer, Heidelberg (2004)

[Milo et al(2002)Milo, Shen-Orr, Itzkovitz, Kashtan, Chklovskii, and Alon] Milo, R., Shen-Orr, S., Itzkovitz, S., Kashtan, N., Chklovskii, D., Alon, U.: Network motifs: Simple building blocks of complex networks. Science 298(5594), 824–827 (2002), doi:10.1126/science.298.5594.824

[Mitchell(1998)] Mitchell, M.: An Introduction to Genetic Algorithms. MIT Press, Cambridge (1998)

[Mjolsness et al(1991)Mjolsness, Sharp, and Reinitz] Mjolsness, E., Sharp, D.H., Reinitz, J.: A connectionist model of development. Journal of Theoretical Biology 152(4), 429–453 (1991)

[Morgan et al(1915)Morgan, Sturtevant, Muller, and Bridges] Morgan, T.H., Sturtevant, A.H., Muller, H.J., Bridges, C.B.: The Mechanism of Mendelian Heredity. Holt, New York (1915)

[Nehaniv(2003)] Nehaniv, C.L.: Evolvability. Biosystems 69(2-3), 77–81 (2003)

[Nehaniv(2005)] Nehaniv, C.L.: Self-Replication, Evolvability and Asynchronicity in Stochastic Worlds. In: Lupanov, O.B., Kasim-Zade, O.M., Chaskin, A.V., Steinhöfel, K. (eds.) SAGA 2005. LNCS, vol. 3777, pp. 126–169. Springer, Heidelberg (2005), http://homepages.feis.herts.ac.uk/~comqcln/nehaniv-SAGA05-withcorrections.pdf

[Nolfi and Parisi(1995)] Nolfi, S., Parisi, D.: Genotypes for neural networks. In: Arbib, M.A. (ed.) The Handbook of Brain Theory and Neural Networks, pp. 431–434. MIT Press, Cambridge (1995)

[van Noort et al(2004)van Noort, Snel, and Huynen] van Noort, V., Snel, B., Huynen, M.: The yeast coexpression network has a small-world, scale-free architecture and can be explained by a simple model. EMBO Reports 5(3), 280–284 (2004)

[Ohno(1970)] Ohno, S.: Evolution by Gene Duplication. Springer (1970)

[Prum and Brush(2002)] Prum, R.O., Brush, A.H.: The Evolutionary Origin and Diservication of Feathers. Quaterly Review of Biology 77, 125–129 (2002)

[Ptashne(1992)] Ptashne, M.: A Genetic Switch, 2nd edn. Cell Press and Blackwell Science (1992)

[Quick et al(2003)Quick, Nehaniv, Dautenhahn, and Roberts] Quick, T., Nehaniv, C.L., Dautenhahn, K., Roberts, G.: Evolving Embodied Genetic Regulatory Network-Driven Control Systems. In: Banzhaf, W., Ziegler, J., Christaller, T., Dittrich, P., Kim, J.T. (eds.) ECAL 2003. LNCS (LNAI), vol. 2801, pp. 266–277. Springer, Heidelberg (2003)

[Rechenberg(1971)] Rechenberg, I.: Evolutionsstrategie: Optimierung technischer systeme nach prinzipien der biologischen evolution. PhD thesis. TU Berlin (1971)

[Reil(1999)] Reil, T.: Dynamics of Gene Expression in an Artificial Genome – Implications for Biological and Artificial Ontogeny. In: Floreano, D., Mondada, F. (eds.) ECAL 1999. LNCS (LNAI), vol. 1674, pp. 457–466. Springer, Heidelberg (1999)

[Sammeth et al(2008)Sammeth, Foissac, and Guigo] Sammeth, M., Foissac, S., Guigo, R.: A general definition and nomenclature for alternative splicing events. PLoS Comput. Biol. 4(8), e1000, 147 (2008), doi:10.1371/journal.pcbi.1000147

[Schlitt and Brazma(2007)] Schlitt, T., Brazma, A.: Current approaches to gene regulatory network modelling. BMC Bioinformatics 8(suppl. 6) (2007), doi:10.1186/1471-2105-8-S6-S9

[Sims(1994)] Sims, K.: Evolving virtual creatures. In: Proceedings of SIGGRAPH, pp. 15–22. ACM Press (1994)

[Stanley and Miikkulainen(2003)] Stanley, K.O., Miikkulainen, R.: A taxonomy for artificial embryogeny. Artif. Life 9(2), 93–130 (2003)

[Stein(2004)] Stein, L.D.: Human genome: end of the beginning. Nature 431(7011), 915–916 (2004)

[Steiner et al(2006)Steiner, Olhofer, and Sendhoff] Steiner, T., Olhofer, M., Sendhoff, B.: Towards shape and structure optimization with evolutionary development. In: Artificial Life X: Proceedings of the Tenth International Conference on the Simulation and Synthesis of Living Systems, pp. 70–76. MIT Press (2006)

[Strogatz(1994)] Strogatz, S.H.: Nonlinear Dynamics and Chaos. Perseus Books (1994)

[Taylor and Raes(2005)] Taylor, J.S., Raes, J.: Small-scale gene duplications. In: Gregory, T.R. (ed.) The Evolution of the Genome. Elsevier Academic Press (2005)

[Taylor(2004)] Taylor, T.: A Genetic Regulatory Network-Inspired Real-Time Controller for a Group of Underwater Robots. In: Intelligent Autonomous Systems, vol. 8, pp. 403–412. IOS Press (2004)

[Turing(1952)] Turing, A.M.: The chemical basis of morphogenesis. Philosophical Transactions of the Royal Society of London Series B, Biological Sciences 237(641), 37–72 (1952)

[Tyson et al(2008)Tyson, Albert, Goldbeter, Ruoff, and Sible] Tyson, J.J., Albert, R., Goldbeter, A., Ruoff, P., Sible, J.: Biological switches and clocks. Journal of the Royal Society Interface 5(suppl. 1), S1–S8 (2008),
doi: http://dx.doi.org/10.1098/rsif.2008.0179.focus

[Vohradsky(2001)] Vohradsky, J.: Neural network model of gene expression. The FASEB Journal 15, 846–854 (2001)

[Waddington(1942)] Waddington, C.H.: Canalisation of development and the inheritance of acquired characters. Nature (150), 563–565 (1942)

[Waddington(1953)] Waddington, C.H.: Genetic assimilation of an acquired character. Evolution 7(2), 118–126 (1953)

[Wagner(1996)] Wagner, G.P.: Homologues, natural kinds, and the evolution of modularity. Am Zool 36, 36–43 (1996)

[Wagner and Altenberg(1996)] Wagner, G.P., Altenberg, L.: Complex adaptations and the evolution of evolvability. Evolution 50, 967–976 (1996)

[West et al(2002)West, Griffith, and Greenberg] West, A.E., Griffith, E.C., Greenberg, M.E.: Regulation of transcription factors by neuronal activity. Nat. Rev. Neurosci. 3(12), 921–931 (2002)

[West-Eberhard(2003)] West-Eberhard, M.J.: Developmental Plasticity and Evolution. Oxford University Press (2003)

[Winfree(1980)] Winfree, A.T.: The Geometry of Biological Time, Biomathematics, vol. 8. Springer, Berlin (1980)

[Winfree(1986)] Winfree, A.T.: The Timing of Biological Clocks. Scientific American Books, Inc. (1986)

[Wolpert(1969)] Wolpert, L.: Positional information and the spatial pattern of cellular differentiation. Journal of Theoretical Biology 25, 1–47 (1969)

[Wolpert(1996)] Wolpert, L.: One hundred years of positional information. Trends in Genetics 12, 359–364 (1996)

[Wolpert and Szathmáry(2002)] Wolpert, L., Szathmáry, E.: Multicellularity: Evolution and the egg. Nature 420(6917), 745 (2002)

[Wolpert et al(2007)Wolpert, Jessell, Lawrence, Meyerowitz, Robertson, and Smith] Wolpert, L., Jessell, T., Lawrence, P., Meyerowitz, E., Robertson, E., Smith, J.: Principles of Development, 3rd edn. Oxford University Press, Oxford (2007)

[Wright(1932)] Wright, S.: The roles of mutation, inbreeding, crossbreeding and selection in evolution. In: Proceedings of the Sixth International Congress on Genetics, Chicago, Illinois, pp. 356–366 (1932)

[Zajac et al(2003)Zajac, Jones, and Glazier] Zajac, M., Jones, G.L., Glazier, J.A.: Simulating convergent extension by way of anisotropic differential adhesion. Journal of Theoretical Biology 222(2), 247–259 (2003), doi:10.1016/S0022-5193(03)00033-X

DATE DUE